Praise for
ein and *Good Without God*

"A passionate introduction to the philosophy of Humanism. . . . Epstein's convivial argument gets beyond the hairsplitting, condescension and animosity of so-called New Atheists like Richard Dawkins, Christopher Hitchens and Sam Harris to arrive at a constructive ideology that explains why it's important to be good even without the presence of the Almighty, and how to do it. . . . An effective primer on Humanism. . . . A timely manifesto for a misunderstood and maligned school of thought." —*Kirkus Reviews*

"An updated defense of Humanism in response to the belligerent attacks on religion put forward by such new atheists as Richard Dawkins, Sam Harris and Christopher Hitchens. . . . [Epstein's] most impassioned argument is with megachurch pastor Rick Warren and other evangelicals who believe secularism is the enemy and a moral society impossible without a belief in God." —*Publishers Weekly*

"Epstein's vision is highly inclusive and emphasizes the vast common ground between the religious and nonreligious without diminishing or compromising the obvious differences. . . . [A] passionate collection of thoughts and ideas." —*Library Journal*

"[Epstein] persuasively claims that the Humanist approach to life can provide the nonreligious with purpose and dignity. A thoughtful account of an often contentious topic." —*Booklist*

"*Good Without God* is not just a brilliant book title but an indispensable and humane ideal. Greg Epstein's wise and warm explanation of the humanist world view goes beyond the recent atheist bestsellers and speaks to the moral and spiritual impulses that have traditionally attracted people to religion."

—Steven Pinker, bestselling author of *The Language Instinct* and *How the Mind Works*

"In *Good Without God*, Greg Epstein shows us what it means to cross boundaries, build bridges, and work toward a society where people from all backgrounds live in equal dignity and mutual loyalty."

—Eboo Patel, founder, Interfaith Youth Core, and member of the Advisory Council of the White House Office of Faith-Based and Neighborhood Partnerships

"Simply the best short introduction to unbelief I've seen."

—Tom Flynn, editor, *The New Encyclopedia of Unbelief* and Executive Director of the Council for Secular Humanism

"[Epstein is] perhaps the most outspoken voice for Humanism in the United States."

—*New York Magazine*

"[Epstein has] probably done as much for Humanism in the U.S. as [Richard] Dawkins has done for it in the UK."

—*The Guardian*

"So wide-ranging is Epstein's rhetorical net, and so agreeable its entanglements, that readers of all stripes will be hard pressed *not* to join the chorus by the book's close: atheist-bashing is shameful! Humanists and believers can get along! Religion can teach atheism a thing or two!"

—*Boston Phoenix*

"[Epstein's] message is direct, while also being compassionate [and] real."

—*Harvard Gazette*

"[Epstein is] one of the [Humanist] movement's most promising young leaders."

—*Jewish Daily Forward*

Rosa Blumenfeld

About the Author

Greg M. Epstein holds a B.A. in religion and Chinese, as well as an M.A. in Judaic studies from the University of Michigan and an M.A. in theological studies from the Harvard Divinity School. In addition to serving as the Humanist chaplain at Harvard University, he is a regular contributor to "On Faith," an online forum on religion produced by *Newsweek* and the *Washington Post*, and his work has been featured on National Public Radio and in several national publications.

Good Without God

What a Billion Nonreligious People *Do* Believe

Greg M. Epstein

HUMANIST CHAPLAIN AT HARVARD UNIVERSITY

HARPER

NEW YORK • LONDON • TORONTO • SYDNEY

HARPER

A hardcover edition of this book was published in 2009 by William Morrow, an imprint of HarperCollins Publishers.

HarperCollins books may be purchased for educational, business, or sales promotional use. For information, please e-mail the Special Markets Department at SPsales@harpercollins.com.

FIRST HARPER PAPERBACK PUBLISHED 2010.

Designed by Richard Oriolo

The Library of Congress has catalogued the hardcover edition as follows:

Epstein, Greg M.
Good without God : what a billion nonreligious people do believe / Greg M. Epstein.
p. cm.
Includes bibliographical references.
ISBN-13: 978-0-06-167011-4
ISBN-10: 0-06-167011-1
1. Good and evil. 2. Virtue. I. Title.

BJ1401.E67 2009
171'2—dc22

ISBN 978-0-06-167012-1 (pbk.)

HB 12.20.2021

Dedicated to Rabbi Sherwin Theodore Wine, 1928–2007

Where is my light?

My light is in me.

Where is my hope?

My hope is in me.

Where is my strength?

My strength is in me . . . and in you.

Sherwin Wine, "Where Is My Light?" (song of Humanistic Judaism)

Contents

Introduction ix

1 Can We Be Good Without God? 1

2 A Brief History of Goodness Without God, or a Short Campus Tour of the University of Humanism 38

3 *Why* Be Good Without a God? Purpose and *The Plague* 61

4 Good Without God: A How-To Guide to the Ethics of Humanism 104

5 Pluralism: Can You Be Good *with God*? 151

6 Good Without God in Community: The Heart of Humanism 169

Postscript: Humanism and Its Aspirations 221
Appendix: Humanist and Secular Resources 227
Acknowledgments 241
Notes 243

Introduction

Can You Be Good Without God?

This is a book about Humanism. If you're not familiar with the word *Humanism*, it is, in short, goodness without God. This is a book about the values, the history, and the future of the world's hundreds of millions of atheists, agnostics, and nonreligious people.

This is not a book about *whether* one can be good without God, because that question does not need to be answered—it needs to be rejected outright. To suggest that one *can't* be good without belief in God is not just an opinion, a mere curious musing—it is a prejudice. It may even be discrimination. After all, would you ever ask: Is it possible to be a good person if you're Muslim? Or Buddhist? Or Jewish? Or Christian? Would you feel comfortable working for an employer who implied that all gays and lesbians were immoral? Or all Democrats? Or all Republicans? How would you feel if your daughter were planning to marry someone who claimed that all Catholics were lousy, unethical human beings? Or all Protestants? This is the sort of all-or-nothing condemnation of a huge population one is making if one suggests that goodness and morality require belief in a deity.

And it's hardly a hypothetical suggestion: over decades of polling, a majority of Americans have consistently indicated a negative opinion of atheists and nonbelievers. Even in this enlightened twenty-first century, where we've proved ourselves ready for a black president and welcomed elected officials representing every group, approximately *half* of all Americans say they would refuse to vote for a well-qualified atheist candidate for public office. In other words, one out of every two Americans admits to being prejudiced against fellow citizens who don't believe in God. No other minority group in this country is rejected by such large numbers.[1]

This prejudice ought to concern us all. Because prejudice anywhere endangers not only its targets, but all who believe that we should be judged not by the color of our skin, or our gender, or sexuality, or by our religious preference or lack thereof, but by the content of our character. If we can convince ourselves today that one entire group comprising millions of people might be incapable of goodness, might be "no good," then we harbor inside us the ability to turn against and hate any other group as well, and no one should feel safe.

It is not easy to live a good life or be a good person—with or without a god. The fact is that life is hard. Living well and being a good person are difficult to do. But that doesn't mean we should give ourselves permission to judge an entire group of people as incapable of goodness unless they're being good the majority's way.

Tolerant, fair-minded people of all religions or none do not dwell on the question of *whether* we can be good without God. The answer is yes. Period. Millions and millions of people are, every day. However, the question *why* we can be good without God is much more relevant and interesting. And the question of *how* we can be good without God is absolutely crucial. Those are the questions in this book—the essential questions asked and answered by Humanism. I invite you to explore these questions, and Humanism's answers, with me.

Are You Religious?

If you're not religious—if you don't believe in God, you're not sure you believe in God, or if you think you believe in some kind of higher power but you know you don't fit into any organized religion—you're not alone. Here's the good news: over a billion people around the world today are like you. All the major studies of world religious demographics, despite different methodologies, indicate that there are somewhere around one billion people on earth who define themselves as atheist, agnostic, or nonreligious. Even if we exclude the approximately half of nonreligious people who say they believe in some form of "spirit"—though I think it makes sense to include many of them—there are still more than half a billion people in the world who live without belief in God. And even in the United States of America, which we're told is the most religious of all the world's developed nations, the nonreligious now represent approximately 15 percent of the population, or approximately 40 million Americans. "Nonreligious" is the fastest growing "religious preference" in the United States, and the only one to have increased its percentage of the population in every one of the fifty states over the past generation. Almost one in four American young adults today has no religion, which suggests not only a growing trend but also that an even larger percentage of the United States as a whole may be secular in another generation. Granted, when pollsters ask Americans to identify themselves as atheists or agnostics, only a few million answer affirmatively. But those terms are attached to a stigma. When poll questions ask in a more roundabout way, such as "Do you believe in God?" the number who say "no" or "not sure" is much higher. And the number of Americans who don't expect to have a religious funeral is in the stratosphere—nearly a quarter of us.[2]

What's more, there's plenty of evidence to suggest that nonreligious people *are* being good en masse. It has long been known that Humanists and nonreligious people have made extraordinary contributions to science and philosophy as well as to philanthropy and social justice. But sociologists have recently begun to pay more attention to the fact that some of the world's most secular countries, such as those in Scandinavia, are among the least violent, best educated, and most likely to care for the poor.[3] And as scientists are now beginning to document, though religion may have benefits for the brain, so may secularism and Humanism. Atheists meditating on positive secular images can gain the same benefits that religious people do from prayer. Strongly convinced nonbelievers may be among the least depressed people—along with strongly convinced believers. Nonreligious

Americans have even been shown to be far more likely than regular churchgoing believers to oppose U.S. government–sponsored torture or "advanced interrogation techniques."

Some say that all these people have nothing in common beyond their nonbelief—or that, because they don't call themselves by the same names or join the same organizations, we should not count or study them. This is nonsense. After all, Christianity is an incredibly diverse tradition as well, encompassing beliefs, customs, and organizations that range widely, from archliberal Unitarian Universalists in Cambridge, Massachusetts, to African American Baptists in Montgomery, Alabama, to Mormons in Salt Lake City, and far beyond. If we study Christianity as a big-tent tradition, or Hinduism (with its thousands of gods and traditions, which many of its followers have trouble agreeing upon), we have to study the nonreligious together as well. We may be a diverse group, but no more so than others.

Still, up to now, only a small percentage of so-called nonbelievers have seen themselves as part of a bigger group of like-minded people, let alone a movement capable of improving people's perceptions of them or making the world a better place.

Are You a Humanist?

If you identify as an atheist, agnostic, freethinker, rationalist, skeptic, cynic, secular humanist, naturalist, or deist; as spiritual, apathetic, nonreligious, "nothing"; or any other irreligious descriptive, you could probably count yourself what I call a Humanist. Feel free to use whatever terminology you prefer—that's not important. We don't believe a god created perfect religions or sacred texts, so why would we believe he or she created one perfect, sacred name that all doubters were required to adopt? And as we've seen in recent years with the success of the GLBT movement—or is it LGBT? Or gay? Or queer?—it's not necessary to reach universal agreement about nomenclature in order to bring a massive group together to gain influence and recognition. The point is that as a Humanist, you'd be in distinguished company, along with Thomas Jefferson, John Lennon, Winston Churchill, Margaret Sanger, Jean-Paul Sartre, Voltaire, David Hume, Salman Rushdie, Elizabeth Cady Stanton, Confucius, Muhammad Ali Jinnah, Wole Soyinka,

Kurt Vonnegut, Zora Neale Hurston, Mark Twain, Margaret Meade, Bill Gates, Warren Buffett, Einstein, Darwin, and more than a billion people worldwide.

All this makes you and me adherents of one of the four largest lifestances on earth, along with Christianity, Islam, and Hinduism. But if we are adherents, what is it that we adhere to? What, if anything, do we have in common? Do the diverse and often disparate multitudes so often dismissed as mere "nonbelievers" share any beliefs in common? Now that we are beginning to gain recognition—such as a positive mention in President Obama's inaugural address or a story about us on the front page of the *New York Times*—it's time to recognize that nonbelievers are believers too: we believe in Humanism.

What Is Humanism?

Humanism is a bold, resolute response to the fact that being a human being is lonely and frightening. We Humanists take one look at a world in which the lives of thousands of innocent children are ripped away every year by hurricanes, earthquakes, and other "acts of God," not to mention the thousand other fundamental injustices of life, and we conclude that if the universe we live in does not have competent moral management, then so be it: we must become the superintendents of our own lives. Humanism means taking charge of the often lousy world around us and working to shape it into a better place, though we know we cannot ever finish the task.

In short, Humanism is being good without God. It is above all an affirmation of the greatest common value we human beings have: the desire to live with dignity, to be "good." But Humanism is also a warning that we cannot afford to wait until tomorrow or until the next life to be good, because today—the short journey we get from birth to death, womb to tomb—is all we have. Humanism rejects dependence on faith, the supernatural, divine texts, resurrection, reincarnation, or anything else for which we have no evidence. To put it another way, Humanists believe in life *before* death.

More formally, the American Humanist Association defines Humanism as a progressive lifestance that, without supernaturalism, affirms our ability and responsibility to lead ethical lives of personal fulfillment, aspiring to the greater good of humanity. This approach, though affirmed by most of the world's hundreds of millions of atheists and agnostics, is not particularly organized. And yet it can be, and it already is in many places, though some secularists bristle at the thought that this is too much like an "organized religion." As we'll see, Humanism is a cohesive world *movement* based on the creation of good lives and communities, without God.

Nonreligious people often wonder why on earth, with all the abuses and scandals and illogical ideas religion is responsible for, is religion still so powerful? The answer is that for most, religion is not about belief in an all-seeing deity with a baritone voice and a flowing beard. It is about group identification—the community and the connections we need to live. It is about family, tradition, consolation, ethics, memories, music, art, architecture, and much more. These things are all good, and no one wants to or should be asked to give them up because of lack of belief in a god.

The truth is that at the present time, the above list of social goods—family, tradition, memories, music, etc.—is difficult to find communally outside traditional forms of religious affiliation or custom. And in truth, being a good person in a vacuum is not a very satisfying experience. Those of us who don't want to worship an invisible being or spend our days fretting about punishment in Hades do want to be able to share what we hold dear with our families and the broader world, and we want to be understood and appreciated for who we are. To do so we need community.

At the most important times of our lives—when we or our loved ones are sick and dying; when a new baby is born; when we want to affirm our love in marriage; when we want to educate our children not only about facts and dates, but also important values—we need to be part of a group. We need what, at least potentially, can be found or created in a *Humanist* community: a place where family, memory, ethical values, and the uplifting of the human spirit can come together with intellectual honesty, and without a god.

Is Humanism a Faith?

I'm often asked whether Humanism is a religion. Practically speaking, Humanism is not a religion, because most of us associate the word *religion* with a system that includes divinities and the supernatural. Humanists have no popes and no perfect people—as the intentionally silly T-shirts say, we are a "nonprophet organization." Sociologically speaking, however, Humanism is similar to a religion in the way that it involves shared values with efforts to organize a community and is essentially a way of life. So I prefer the European term *lifestance*, meaning more than a philosophy but not a divine or revealed religion. In any case, asking whether Humanism is a religion or not is little more than a semantic "gotcha" game. Some ask because they're religious and are trying to knock Humanists down a peg; others are angry at religious hypocrisy and are afraid to be associated with anything that even faintly smells of *belief*, a word laden with baggage and unhappiness.

But the point is not whether you believe in something, it's what you believe *in*. Not believing in anything is a belief too—in *nihilism*. As TV and film writer/director Joss Whedon said, "The enemy of Humanism is not faith—the enemy of Humanism is hate, it is fear, it is ignorance, it is the darker part of man that is in every Humanist, and every person in the world . . . But faith is something we have to embrace. Faith in God means believing absolutely in something, with no proof whatsoever. Faith in humanity means believing absolutely in something with a huge amount of proof to the contrary. We are the true believers."[4]

How Is Humanism Different from the New Atheism?

So much has been written about the religious people and traditions of the world. Thousands of anthropologists and sociologists have devoted their lives to studying religious traditions and their adherents. Millions and millions of pages have been written about theology—about what religious people believe. But try to go to your local bookstore or library and ask for a book about nonreligious people or what we believe. The choices have always been scant indeed. So it's no wonder the recent spate of best-selling books by atheists attacking religion has caused such a stir.

Today, those who believe that the good life ought to be defined as obedience to God and tradition feel under siege by the forces of modernity. In their minds, certain outward signs of this modernity—whether gay pride paraders all done up in leather and fuchsia, a woman rearing a child on her own, or simply people like me who can publicly deny a belief in God and live respectable lives—are all declarations of war against the old ways. And so both fundamentalist Islam and fundamentalist Christianity, among other religious forces, have declared war on secularism and Humanism. In turn, a group of bold new atheist intellectuals and leaders has arisen to declare war right back, proclaiming "God is not great!" "God is a delusion!" and "This is the end of faith!"

I admire today's "new atheists" because they seek to right the very real and very many religious wrongs of our time. And I especially appreciate Messrs. Dawkins, Harris, and Hitchens when they liberate young people to feel good and be open about their lack of belief in God at a time when many still live in communities that shun those who will not produce at least an outward display of allegiance to the old values. But atheism goes astray when it adopts a certain posture, one best captured by a cover story in *Wired* magazine in November 2006: "The New Atheism: No Heaven. No Hell. Just Science."

It is true and important that Humanists don't adhere to the idea of a heaven or a hell, and it is also true that we value science as the best tool humans have for understanding the world around us. But "Just Science"? Such language raises concern that the new atheism is cut off from emotion, from intuition, and from a spirit of generosity toward those who see the world differently. While nonreligious people often value science highly, many deeply religious people value and study it as well. So surely valuing science cannot be a way to distinguish religious people from nonreligious people. Besides, books on science, though often containing much useful information about the world around us, can less often say important things about what we ought to value most in life, or why. Science can teach us a great deal, like what medicine to give to patients in a hospital. But science won't come and visit us in the hospital.

This may seem like just a cute play on words, but when I was a young boy in a nonreligious family, I had to spend a fair amount of time in hospitals, and they were often lonely places for me and my family. When I was eighteen and my father finally died after battling lung cancer for years, I knew he missed the religious communities he had left behind earlier in life. Though he was not exactly the greatest at communicating his feelings of isolation, you could tell he longed for community because when the Jehovah's Witnesses would knock on our door to try to convert him, he would always drag himself out of his sickbed, a little of his gloom lifting at the opportunity for companionship and serious conversation with new people. Still, the young men did not come often, and in any case, given their wildly different worldview from his own, they were not ideal companions. So my father died feeling quite alone, never having heard of the idea of Humanist community.

I myself, despite having majored in religion in college, only learned of Humanism after graduation, because of a chance run-in with a great Humanist leader, the late Sherwin Wine. Sherwin, to whom this book is dedicated, was a Humanist rabbi (more on what that is later) who visited thousands of congregants over his long career as a clergyman who believed—openly—in good, not God. It was Sherwin who first answered many of my early questions about Humanism—the basic questions that are the subject of the first chapter of this book, "Can We Be Good Without God?"

The history of goodness without God as a world tradition has roots in the ancient East and West. Humanism traces its story back not only to the European Enlightenment and to ancient Greece, as many assume, but also touches cultures from India, China, and the Middle East. It is a belief held by American Revolutionary patriots like Thomas Jefferson, leading women's suffragists of the nineteenth century, civil rights leaders of the twentieth, and on to the original new atheists—not only Dawkins, Hitchens, Dennett, and Harris but Freud, Marx, Nietzsche, and Darwin.

Chapter three is an exploration of the question *why* are we good without a god? Religion is a profound source of meaning and purpose for many people—even for those who, despite more than a little bit of skepticism about its supernatural claims, fear that without religion there is simply no reason to live, or at least no reason to live morally and ethically. But a Humanistic approach to life can provide nonreligious people with a profound and sustaining sense that, though there is no single, overarching purpose given to us from on high, we can and must live our lives for a purpose well beyond ourselves.

But it's not enough to just "discover" the meaning of life. What really matters is whether we live according to our values, and that takes hard work and a hundred hard choices every day. What is good without God? From learning how to put the golden rule and human compassion into action more often, to exploring innovative and sustainable answers to new human problems such as climate change and other bioethical dilemmas, Humanism is concerned with one of the most important ethical questions—what we *do* once we've found purpose in life.

It's also important to acknowledge people who consider themselves religious. Welcome. I hope Christians, Muslims, Jews, Buddhists, Hindus, Sikhs, Baha'is, and many others are reading, and will read on—not to be converted, or deconverted, but to gain understanding of loved ones and neighbors who may be Humanists. I want to offer an affirmative response to the question can you be good *with* God? I urge atheists and agnostics to strive for what Steven Prothero calls religious literacy, and I implore religious people and Humanists to enter into deeper dialogue and cooperation—because we live in a world that is flat, interconnected, interdependent, not to mention armed to the teeth with weapons of mass destruction—a world where we can no longer afford to misunderstand one another or to be ignorant about what makes each other tick.

I believe that community is the heart of Humanism. In the past century, God was supposed to be dead, but too often it has seemed that Humanism died instead. What will it take for a new Humanism to arise—one that is diverse, inclusive, inspiring, and a transformative force in the world today?

If this book accomplishes one thing for or on behalf of the billion nonreligious people, let it not be that we learn how better to convince others that there is no God, or that religion is evil. May we encourage more hospital visits by the nonreligious, both literally and metaphorically. May we do more good work together and build something positive in this world—the only world we will ever have. May we focus more on the "good" than on the "without God."

Good Without God

CHAPTER ONE

Can We Be Good Without God?

So It's Been Said, Many Times, Many Ways: How Should You Respond When Your Way of Life Is Insulted by One of the Most Powerful Men in the World?

There is an ethical dilemma I have struggled to find the right answer to—because to be honest, one of my deepest convictions as a Humanist is that no one person can ever have all the answers. If you ever meet anyone who tells you his or her religion can offer all the answers, run for the hills. Or at least hide your wallet.

But the dilemma I'm talking about is one that millions of people face every day: how should you respond when your way of life is insulted by one of the most powerful men in the world?

This is not necessarily a life-or-death question. Though it could be. In 2006 an obscure cartoonist for a Danish newspaper drew a series of cartoons depicting the prophet Muhammad as a comically angry terrorist. Hilarity did not ensue. Hundreds of thousands rioted. Embassies burned.

A number of years earlier, novelist Salman Rushdie satirized the process of writing the Qur'an in his *The Satanic Verses* and nearly paid with his head after receiving an "unfunny valentine" (Rushdie's words) from the Ayatollah Khomeini of Iran, a fatwa calling for his death. And so if cartoonists and novelists can provoke such a reaction, what happens if such insults are made by someone truly influential? Like, say, the most powerful preacher in the most powerful country in the world—a megachurch minister who is said to have trained over four hundred thousand evangelists for leadership in Christian communities and who, after a long summer in 2008 when it seemed that Barack Obama and John McCain would never appear together, was the first to bring the Democratic and Republican presidential nominees together publicly so that each could be interviewed *by him and him alone*?

But I'm not asking what Muslims would do if the Reverend Rick Warren were to insult them. I'm wondering how the nonreligious should respond to his repeated statements that he could never vote for an atheist as president of the United States—including saying so on *Larry King Live* just hours after hosting his Saddleback Forum with Obama and McCain in August 2008. Warren has in fact become this generation's most eloquent, charismatic advocate for a certain very old opinion that lashes out against the worth and dignity of those who consider themselves atheists, agnostics, or nonreligious. In *The Purpose Driven Life*, he writes:

> If your time on earth were all there is to your life, I would suggest you start living it up immediately. You could forget about being good and ethical, and you wouldn't have to worry about any consequences of your actions. You could indulge yourself in total self-centeredness because your actions would have no long-term repercussions. But—*and this makes all the difference*—death is not the end of you! Death is not your termination, but your transition into eternity, so there are eternal consequences to everything you do on earth. Every act of our lives strikes some chord that will vibrate in eternity.[1]

Warren's tone here is positive, even inspiring, despite the negative message that a nonreligious person, one without a *literal* belief in resurrection, has no reason to be decent. So lest he be misunderstood as more broad-minded and tolerant than he is, let's note that on the previous page of *The Purpose Driven Life,* he is quite clear about what he thinks of nonbelievers: "While life on earth offers many choices, eternity offers only two: heaven or hell. Your relationship to God on earth will determine your relationship to him in eternity. If you learn to love and trust God's Son, Jesus, you will be invited to spend the rest of eternity with him. On the other hand, if you reject his love, forgiveness, and salvation, you will spend eternity apart from God forever."[2]

Here, despite Warren's eloquence, the author of a book billed as the best-selling nonfiction book in history (one might ask, does this imply that the Bible is *fiction?*) is damning you to eternal hell if you are not a traditional Christian. It's a profound insult—maybe against you, maybe against your mother, father, spouse, child, or friend.

But how to respond? Should we insult Warren back, call him a buffoon, a charlatan, a brainwasher for Jesus—take the rhetorical eye for an eye? Or should we turn the other cheek? Should we ignore him? I was taught as a child that it is best to ignore petty schoolyard taunting. But is it really safe to ignore such a deep-cutting insult from a man who is famous not merely for being pretty, or athletic; but for being wildly *influential*?

It would be one thing if it were only Warren who was leveling this charge that we can't be good without God. But plenty of other influential Christians have been doing so for a very long time now—for example, C. S. Lewis, with his all-too-common claim that when we debunk God, we become slaves to our base impulses, left with no ethical foundation: "When all that says 'it is good' has been debunked, what says 'I want' remains . . . The Conditioners, therefore, must come to be motivated simply by their own pleasure . . . My point is that those who stand outside all judgments of value cannot have any ground for preferring one of their own impulses to another except the emotional strength of that impulse."[3]

Other Christians like to emphasize that if we lose God, we lose "absolute values," without which we will end up in the moral gutter. Albert Mohler, head of the influential Southern Baptist Theological Seminary, gives this frequent insult concise and charismatic voice on his blog: "If human beings are left to our own devices and limited to our own wisdom, we will invent whatever model of 'good character' seems right at the time. Without God there are no moral absolutes. Without moral absolutes, there is no authentic knowledge of right and wrong."[4]

But it is not only Christians who say we can't be good without God.

In the Muslim world, perhaps the most powerful and portentous theological message of the past half century has been the notion that secularism, whether in the West or the Middle East, is the corrupt and degenerate "reign of man." "Humanity today stands on the brink of the abyss . . . To establish the reign of God on earth and eliminate the reign of man, to take power out of those who have usurped it and return it to God alone." This is a quote from not just any Muslim thinker but Sayyed Qutb, aka Osama Bin Laden's ideological godfather, who added chillingly and prophetically, "this will not be done through sermons and discourse."[5]

And while you may not be shocked to hear of conservative Muslim thinkers agreeing with conservative Christians in their denigration of atheism and secularism, I was surprised to note—as a graduate student several years ago taking a class on liberal, progressive, and pluralistic voices among the world's religious traditions with Diana Eck, founder of the Pluralism Project and one of America's foremost advocates for interreligious understanding—that even many of the world's most open-minded Muslim theologians see secularists and atheists as villains to be rallied against. In the words of Seyyed Hossein Nasr, a Harvard- and MIT-trained professor of Islamic studies and author of *The Heart of Islam: Enduring Values for Humanity*, "secularism is the common enemy of all the Abrahamic traditions, and the erosion of moral authority in secular societies that we observe today poses as many problems for Jews and Christians as it does for Muslims." Nasr later continues, "the idea of living at peace while denying God is totally absurd."[6] These words of "wisdom" are from a book Eck assigned at Harvard as an example of progressive, open-minded Islam. (The book has since been removed from Eck's curriculum.) Read liberal Islamic theology today, and unfortunately you'll find a lot more of this approach—attempts to find common ground with Christians and Jews at the expense of demonizing an easy-to-attack, badly misunderstood secular enemy.

And to add yet another voice to the modern chorus calling out that we can't be good without God, *Hitler* agreed with much of what is recorded above. Hitler is often erroneously labeled a secularist or atheist by ignorant religious blowhards desperately searching for a response to the complaint that the Crusades and the Inquisition were religious movements that caused great suffering in the name of God. Of course, the Crusades and the Inquisition *were* religious and did cause great suffering, but beyond acknowledging that here, you won't find me rehashing such attacks on religion. My purpose is not to rake faith in God over the coals in every possible way. Nevertheless, let's be clear that neither Hitler nor Nazism is ever an appropriate comeback to such arguments against religion. In carrying out the Holocaust, Hitler wrote, "I am acting in accordance with the will of the Almighty Creator: by defending myself against the Jew, I am fighting for the work of the Lord."[7] The Nazi army's *belts* were inscribed "Gott mit uns!" (God is with us).

This is what baffles me most about the you-can't-be-good-without-God canard, what makes my jaw drop with disbelief: considering Hitler and bin Laden and others like them, it is probably the only belief that has united some of the most despicable human beings to walk the earth during the twentieth century, and yet we still take the argument seriously, whether it comes from Rick Warren or just some poor lost soul looking for an evil enemy. It saddens me to reflect that such polar opposites in theological, political, and cultural terms find some sick common ground in denigrating my values and worldview. But they do, they are not alone, and their thinking is not new.

The History of a Lousy Idea

Indeed, no man can be good without the help of God.

—Seneca, first century CE

The idea that we can't be "good without God" has been circulating in religious and philosophical literature for a long time. Almost two millennia at least, since the Roman Stoic philosopher Lucius Annaeus Seneca said so in the middle of the first century of the Common Era. And even then, it didn't mean much. When Seneca said it, he wasn't talking about the Christian God, or the Jewish God, neither of which was a blip on the radar screen of most Greco-Roman philosophers at that time. Seneca wasn't even saying we couldn't be moral without a particularly robust version of Zeus or another member of the pagan pantheon. If anything, Seneca's message was liberal and nontraditional, a clever way to argue *against* the kind of zealous god-worship you can find reenacted in the HBO series *Rome*, complete with young women bathed in the blood of a freshly slaughtered cow in the ardent hope that the sacrifice will bring good fortune in a coming battle.

Seneca, a tutor to the young man who would become Rome's Emperor Nero, was writing to a young charge about how, in general, one becomes a good person. Never a task for the fainthearted, and recognizing this, he first established his credibility the only way one really can on such issues: by admitting he didn't have all the answers. Seneca explained, in his *Epistles to Lucilius*, that he asked to be thought of not so much as a doctor come to cure all the patient's illnesses but as a caring fellow patient, recognizing compassionately that he was in a shared hospital room with other frightened patients, and choosing to talk to himself (and his pupil) in the hopes that the thought he put into saving himself from life's errors would benefit others as well.

Seneca began his answer to the question of how goodness, wisdom, and understanding are achieved with the more obvious answer that how we *don't* get them is through praying for them: "We do not need to uplift our hands toward heaven, or to beg the keeper of a temple to let us approach his idol's ear, as if in this way our prayers were more likely to be heard." So far, pretty secular stuff. Seneca continued, however, with a statement that would not only have been considered radical at the time, but is also the kind of theology that gives fundamentalist Christian preachers fits to this day: "God is near you, he is with you, He is within you. This is what I mean, Lucilius: a holy spirit indwells within us, one who marks our good and bad deeds, and is our guardian."

This is God as the inner voice, the combination in popular Christian terms of Santa Claus marking whether we have been naughty or nice and Mother Mary watching out for us when we are alone and troubled. It's the most convenient kind of god, actually—one who takes care of us when we need care and compels neither our subscription to any specific theological creed about him nor tithing to any temple to keep his name up in golden lights. It is about this sort of deity that Seneca then penned the infamous line "Indeed, no man can be good without the help of God." Seneca's statement, then, is really more like "no man is good without being true to himself." He concludes this passage with the statement that there is no external force compelling us to be good—that we must compel ourselves to do so.

A fair question to ask about this whole sordid episode in intellectual history, then, might be what in heaven or earth was Seneca even talking about? Or, why did Seneca feel the need to write about this kind of God? Most likely it was because, given what Seneca believed, he was just the sort of person who in his day might easily have been accused of atheism himself—a punishable sin among the Romans.

A man with Seneca's beliefs—that you couldn't simply sacrifice a virgin to Mars to win a war, or squash a cockroach in honor of Janus, Gaia, and Dis to bring good fortune to your poor family—would have been dangerous in those days not only to those who managed the temple cults but also to the political leaders who justified their often brutal decision-making by claiming the gods' favor. In fact, we religious skeptics have always been a little dangerous to those religious and political authorities who have nothing to offer their people in this world and so must promise more fortune in the next one to maintain control. We continue to be so. But conservative authorities have, since ancient days, had a clever counterstrategy against religious skepticism—convincing people that atheism is evil, and then accusing their enemies of being atheists.

We now come back to the question of how to respond to Rick Warren's insult. It represents a long tradition, which doesn't make the choice of whether to lash back or turn the other cheek any easier.

A large part of the problem is ignorance. What we do not understand frightens us. Fear begets prejudice. And not only do most people have no idea who we nonreligious people are, or what we stand for, we also aren't usually able to articulate much about ourselves and our own beliefs. We know what we *don't* believe. But not what we *do*. And so we become a blank slate, a convenient place for religious people of all kinds to project their fears about immorality and degeneration.

No matter how you think we should respond to these fears and the insults that so often accompany them, a large part of the solution must be education. We must do more to spread awareness about our answers, our beliefs, and our values. As Humanists, nonreligious people have positive, inspiring answers to all of life's great questions, starting with the most basic—who are we, and where do we come from?

In the Beginning

"Fifteen billion years ago, in a great flash, the universe flared forth into being."[8] These are the opening words of *The Universe Story*, by mathematician and cosmologist Brian Swimme and the late cultural historian Thomas Berry, who illustrate how our true origins are even more grand and awe-inspiring than the stories in Genesis or any of the world religion's sacred texts.

We are all part of an amazing story in that, as Swimme and Berry put it, "every living being of Earth is cousin to every other being."[9] Our history began with the Big Bang, a "primordial flaring forth"; it continued with this galaxy's first star, which appeared five billion years later, and the Milky Way's birthing of our sun five billion years ago. With the formation of Earth a billion years later came the first living cell, and then two billion years after that came new kinds of cells that "invented" both sexual reproduction and the predator-prey relationship. These twin developments led to an ever-quickening spiral of change: from the first multicellular animals, to mammals who could sense their environment and feel emotion, to human self-awareness and the ability to stand upright and use tools, to the domestication of fire and the human creation of myth, agriculture, villages, religion, culture, cities, and eventually to the three universalist religions (Buddhism, Christianity, and Islam), mass migration, liberal democracy, the multinational corporation, and *American Idol*.

This is the story of evolution. Humanists accept that the scientific evidence for evolution is overwhelming; we build our worldview around it because we want to look reality square in the face, unblinking, unflinching, unafraid of the truth. But the story of evolution is not the sole property of Humanism. Billions of religious people have accepted the basic tenets of this story as well. The Catholic Church, with its 1.1 billion members, has officially affirmed the reality of evolution—and has acknowledged that "intelligent design" is not science. Of course, I vigorously disagree with what I see as the contorted, nonsensical logic by which Catholic doctrine seeks to suggest that evolution and the traditional Biblical creation story can both be true. But to the extent that Catholics and other religious groups accept that we have reliable evidence for evolution, they are our allies and our friends.

What is unique about the Humanistic view of evolution is that it affirms that the process, from its very beginnings, must be considered unguided. As Richard Dawkins explained in *The Blind Watchmaker,* it is "gloriously and utterly wrong"[10] to assume that living things must have had a conscious, purposeful creator the same way that a pocket watch does. As much as it might seem noble to have been created for a reason, true nobility for us lies in being honest about being able to discern no purpose given to human beings by the Big Bang. All the evidence suggests that creation narratives like that found in Genesis are neither literally true nor divinely inspired metaphors but simply the first flawed human attempts to answer questions for which we now have much better answers. The billions of stars and trillions of big and small rocks that surround us in this universe do not care about us, and do not love us. They do not hear our prayers. The only guidance for which we have ever seen evidence is human guidance. The only purposes we've ever been able to understand are the purposes we have created and chosen. A blind watchmaker created us. But it is now time for us to open our eyes and take responsibility for our future.

How Do We Know Our Story Is True?

Humanism is not in the business of absolute certainties. We do not claim to be able to prove that evolution happened or that there is no God in the same way that we know 2 + 2 = 4. But too many theologians, clerics, and other self-styled religious authorities have tried to convince the credulous that this means there is no difference between the scientific story I just recounted and any traditional faith as they answer the important questions about who we are and how we got here. We should remind them that the question is not whether one believes, but what evidence one's beliefs are *based on.*

All beliefs are based on something. Beliefs in supernatural events such as miracles are based on tradition—such as reading about it in a book we're told is sacred—or intuition—as in those moments when it seems there *must* be something looking out for us. And sometimes traditions and intuitions are correct. But they are not reliable ways to determine whether something is true. Think of it this way: would you want to fly in an airplane designed by an engineer with no advanced scientific degree, who in fact did not believe in science, and instead consulted the Bible or the pope for advice on how to build airplanes? If you developed a potentially fatal disease like cancer, would you want to be treated by a friend with no medical training, who claimed only to have a strong *sense* of how to heal you? Of course you would not. The scientific method, while imperfect, is the most reliable tool human beings have ever known for determining the nature of the world around us.

Call Humanism a *faith* if you like—we should have no particular allergy to that word—but recognize that it is a faith in our ability to live well based on conclusions and convictions reached by empirical testing and free, unfettered rational inquiry. In other words, we question everything, including our own questions, and we search for as many ways as we can to confirm or deny our intuitions. We have no holy books meant to be taken at face value or blindly obeyed. We are open to revising any conclusion we have made if new evidence appears to contradict it.

However, we also recognize that there can and often is a point where sufficient evidence has been gathered on a certain subject to make a reversal of views extraordinarily unlikely, and where the explanation we have pieced together works extremely well. This is certainly the case with whether the sun will rise tomorrow, and it is equally true for evolution, and about our basic picture of the origins of the universe.

Now, sometimes you have people who legitimately do not understand the scientific method—I know I have often felt confused about the process by which scientists gather evidence, test hypotheses, and draw conclusions about, say, the fact that there may be several hundred billion galaxies as opposed to only a few dozen billion. The solution to this problem is education—about science, about what it means to be skeptical of untested claims. Some of our best educators about science in the past generation have also been active, committed Humanists, such as Carl Sagan, Richard Dawkins, E. O. Wilson, and many others. But others who have considered themselves religious have also done excellent scientific work. We disagree with their reasoning on theological issues, but we are capable of working with them extremely well in scientific and public policy endeavors.

As for those who manipulate science and Humanism for their own political and social gain—who intentionally or out of extreme and willful ignorance mock us as having no sound basis for our beliefs—who mislead millions of people into thinking that evolution is a hoax, a joke, a satanic plot—we must engage these individuals forcefully. That doesn't mean by military force, though sometimes Christopher Hitchens sounds as if he *does* want all-out war with the godly, but I digress. Sometimes we need to point out that some of the ways we are attacked and criticized are absurd, even downright funny. Sometimes the most forceful way to respond to our conservative religious critics is with humor.

Bertrand Russell created his "orbiting space teapot" theory to show that yes, we acknowledge that we cannot prove there is *not* a God, or gods, or goddesses, or spirits or ghouls or goblins (it is impossible to prove this kind of a negative, after all), but that we also acknowledge there might be a giant floating vessel of silky Darjeeling tea somewhere off in the far reaches of the galaxy, and we have not a shred of credible *evidence* for the existence of any of them. For this same reason, a brilliant young American named Bobby Henderson recently created the "Church of the Flying Spaghetti Monster." Henderson, nonplussed by the efforts of right-wing Christians to force the teaching of intelligent design in public schools in his home state of Kansas, asserted that *of course there is a designer of the universe*—and that he had been personally touched by the Creator's noodly appendage. Equal time, he demanded, must be given in public schools for the teaching of the spaghetti creation controversy, because Henderson, his many coreligionists, and all their traditional texts asserted plainly that their doctrine was true.

I admit I find the cult of the FSM hysterically funny, but I have no desire to offend my liberal religious allies, and it would be nice if this sort of humor could be restricted in its distribution to those who need to hear it—if our religious allies could be spared a sense that they were being mocked by our secular sensibilities. But that is not the way public discourse works. When James Dobson or Rick Warren makes a claim about the world, not only are conservative Evangelicals in a position to hear it; everyone is. The same goes for the pope, or whoever else. And so, when an atheist or Humanist spokesperson responds with humor to the more unreasonable, indeed ridiculous claims made by some religious authorities, everyone is going to have to hear and interpret it in his or her own way. The point is not to mock religion, but simply to drive home that we have high standards when it comes to deciding whether a story is true or not. Those who want to convince us that there is a God, and that a certain religion has access to eternal truth, should be expected—just as Humanists should be—to produce serious, credible, testable evidence in support of their claims.

Is There a God?
Or, Doesn't It Take Just as Much Faith Not to Believe?

Now, just because science makes good sense—because it can help us fly the plane we want to take us places, or create medicine that can heal us—doesn't mean that the Big Bang seems logical when we first think about it. It's pretty reasonable, given that we don't know what happened before the Big Bang, to wonder, what created it? A lot of reasonable people, therefore, think to themselves that it's just as good as anything else to think God created the Big Bang, and therefore it's just as good as anything else to assume that a lot of other things religion teaches about God are true too. This is why so many smart people still think that a conversation about what is true and meaningful and worthwhile in life begins with the question "Do you believe in God?"

But "Do you believe in God?" is a totally meaningless question. The real question all people—whether secular or religious—ought to be asking themselves and one another is "What do you believe *about* God?"

Why the seemingly small distinction? Because in this day and age, the word *God* can mean anything; and so it means nothing.

Recently I had a conversation with a student I'll call Jennifer, who was interested in the Humanist community at Harvard but was clearly struggling with the idea of coming out and saying she *didn't believe in God.* I felt for her: she had a strained, painful relationship with her conservative father, to whom she longed to be closer, and she was afraid that if she said she didn't believe, he'd finally disown her. Still, she's a woman of honesty and integrity—which is why she came out as gay, straining their relationship in the first place. She just couldn't imagine lying about her religious beliefs simply because that was what he wanted to hear. But it was obvious she wanted to be able to say she believed in God, thinking that might smooth things over a little with her family, not to mention give her a sense of security and stability. She didn't want to have to rebel against *everything.* So Jennifer was more than a little bit intrigued, and maybe relieved, when Tim Keller, an Evangelical minister in New York City who specializes in working with educated young people, came to Harvard to read from his new book, *The Reason for God.* Keller's book is meant to demonstrate that there are good philosophical and intellectual reasons for smart people to believe in God. Keller doesn't actually succeed in demonstrating that God exists, but he doesn't need to: all he really has to do is make God seem intellectually respectable for thoughtful people who *want to be able to believe.*

So Jennifer asked me the question, prominently highlighted in Keller's book, that inevitably follows "Do you believe?" for people like her who want their own answer to be yes: "Doesn't it take just as much faith not to believe in God?"

How would you have responded to her?

I knew she was asking partly out of pain and genuine searching, and so I wanted to give her more than just the familiar answer about how we Humanists think scientifically and the "burden of proof" is on those who assert a belief, not those who deny it. I wanted to respect her vulnerability. I had no interest in seeing her brainwashed in either a religious or secular direction, but I also wasn't interested in sugarcoating my response. So I told her to remember that a lot of people say they believe in God, but what they mean by *God* may not be all that relevant to their lives or to the world. The word *God* can refer to a vengeful Old Testament deity; an all-loving New Age Earth-nurturing spirit; or to Love itself, the universe, the "Ground of Being" (whatever that means), or anything else you want it to mean. I encourage you to be as specific as you can possibly be in your own mind about what kind of god you're referring to when answering the question of whether you "believe in God" or not. That way, when you arrive at an answer, there's the greatest chance it will be *your* answer and not Tim Keller's or mine or anyone else's.

In other words, as I mentioned above, the question I wanted Jennifer to ask herself, and the one I'd like you to ask yourself is "What do you believe about God?"

Here is the Humanist answer: we (the nonreligious, atheists, Humanists, etc.) believe that God is the most important, influential literary character human beings have ever created.

This answer is meant seriously, though of course it will satisfy no one. Religious readers will grumble that it is atheistic; atheists will grumble even louder that it avoids the issue, beating around the burning bush of God's nonexistence. But it is also an answer that emphasizes the real point of Humanism, which is that God is beside the point.

Again, Humanists believe that God is the most important, influential literary character human beings have ever created. And it's important to know precisely what we mean by the word *God*, because the more a word can mean anything we want, the more it means absolutely nothing.

A Brief History of the Definition of *God*

When, where, and how did it happen that the word *God* evolved so dramatically? The story of human doubt about fantastical, invisible personal deities is long and deep and wide, and is chronicled elegantly by Jennifer Michael Hecht in her classic work, *Doubt: A History*. Hecht's book traces, almost encyclopedically, the many forms skepticism about religion has taken across the millennia and around the world.

But for now, to answer the persistent questions about what Humanists believe about God, as well as Jennifer's question about whether it might just be easier to believe, we need to look at two key moments in *God*'s coming to mean just about anything in America today—Spinoza's redefinition of the word in the seventeenth century, and John Dewey's continuation of Spinoza's work three centuries later.

Spinoza's God and Tillich's Ultimate Concern

When asked if he believed in God, Einstein famously responded, "I believe in Spinoza's God who reveals himself in the orderly harmony of what exists, not in a God who concerns himself with fates and actions of human beings."[11] So, did Einstein "believe in God"? Did Spinoza? Baruch Spinoza, a Dutch Jewish philosopher in seventeenth-century Amsterdam, was what has since come to be called a pantheist: a person who no longer believed in the

God of the Bible and of traditional Judaism, and who, armed with the new knowledge beginning to emerge at that time about the cosmos and the laws of nature, redefined God as nature and as the universe itself.

At no point did Spinoza make any explicit statement such as "There is no God" or "I do not believe in God." Yet it was understood by almost everyone in both the Jewish and Christian communities surrounding him that by affirming belief in this sort of God, Spinoza declared himself an atheist. Pantheism and atheism were acknowledged to be exact equals, despite the fact that the former provides one with a rhetorical justification for saying "I believe in God" and the latter does not. For this offense of pantheism and atheism (though again, Spinoza did not accept the latter term and lived before the former was coined), Spinoza was permanently banned from his Jewish community—a fate he did not relish, because although he did not believe in the God of his fellow Jews, he did believe in their essential goodness and wanted to be a part of their community. But he was not willing to be dishonest about philosophy or theology in order to live among them, and so he died still in banishment, years later.

Today such matters are handled very differently. Noted modern Christian theologians such as the late Paul Tillich or today's John Shelby Spong can be quite clear that the Christian God in whom they affirm belief possesses no supernatural characteristics whatsoever.[12] They suffer no exile for their heresy. Tillich, who taught at Union Theological Seminary in New York, the University of Chicago, and Harvard in the first half of the twentieth century, and has influenced generations of mainstream liberal political and social leaders, wrote that "faith is the state of being ultimately concerned."[13] Tillich added that because God is to be defined as such, all those who have an ultimate concern, something they "take seriously without any reservation," cannot call themselves atheists or agnostics.

Tillich's is a popular, influential, and widely respected approach to theology, yet he makes the difference between an atheist and a Christian into nothing more than a slippery semantic game.

Surely, "ultimate concern" can be accomplished without belief in a God. Humanists, atheists, and agnostics can be deeply, "ultimately concerned" by their efforts to live with dignity, to treat themselves and others with respect. Just like anyone else, we are "ultimately concerned" with finding a decent, loving way to relate to our families and to the environment. But because we do not call our "ultimate concerns" God, we are treated and seen differently. Why?

Dewey's Reconstructionism vs. Antagonistic Atheism

In the early twentieth century, John Dewey was considered by many as America's foremost philosopher. Born in Burlington, Vermont, in 1859, Dewey was a progressive American patriot who believed in applying the ideas in his pragmatic, Humanistic worldview to better the lives of people across the nation. Dewey believed that providing Americans with better education, including science and the scientific method as our most important tool for determining the truth, would allow them to be better citizen participants in their shared civic life. Dewey took a scientific view of religion and, when approached by a group of nontheistic Humanists who, in the early 1930s, were attempting to draft a declaration of values for those who were good without God, he readily offered his name as a supporter of their "Humanist Manifesto." In fact, he saw his work as a Humanistic alternative for the American masses, an alternative way to envision the religious nature and identity of America. He saw entire public school systems in the United States ultimately adopting the Humanistic value system he helped to shape. This was a big, patriotic goal that would need mainstream approval—but could he tell the masses that Humanism didn't include a God?

Meanwhile, British philosopher Bertrand Russell and others such as Marx, Freud, and Nietzsche were coming to represent another new school of thought that we might call "antagonistic atheism." This was the view that religion was a thing of the past and ought to be brought hastily toward a point of declining influence. These new atheist thinkers in many cases boldly identified with Communism, Socialism, or other radical new ideologies. While Dewey found himself agreeing with many of their insights into metaphysics, he often bristled at their politics and their rebellious, unpatriotic beliefs.

Such concerns were clearly on his mind as Dewey spent the two decades leading up to 1934 largely silent about whether he "believed in God." In other words, he avoided the question. But in 1934 he published his last major work dealing with religion, *A Common Faith*, in which he wrote that one could and should say "God" in reference to all the forces of good that helped humans to live well. Essentially Dewey proposed a formal, lightly refined version of Spinoza's theology. God was not the universe, but the positive forces in the universe as far as humans were concerned. Dewey was not exactly a pantheist, but rather called himself a "Reconstructionist"—one who reconstructs the definition of the word *God* in order to refer to natural human values instead of a supernatural deity.

A Common Faith was well received by general audiences. Humanists largely rejected it. Corliss Lamont published a review entitled "John Dewey Capitulates to God." Dewey's student, the Columbia scholar of philosophy Sidney Hook (later mentor to Paul Kurtz, the most prominent living Humanist philosopher for much of the past generation), told Dewey that the use of God-language was bound to confuse his audience. Dewey initially denied this, but admitted only a few years later that Hook had been at least somewhat correct. A certain kind of clarity about what we mean when we speak of God would never again be achieved in American public life. Dewey's Reconstructionist approach of openly redefining the term *God* has been adopted, consciously or unconsciously, not only by many a skeptical-minded politician hoping not to be "outed" as an atheist the way Spinoza once was, but also by other influential philosophers and theologians such as Mordechai Kaplan, who founded the movement of Reconstructionist Judaism as a combination of Dewean thought and Jewish tradition. Dewey has even trickled down to pop culture tastemakers such as Oprah Winfrey, who has inspired millions of everyday people with the Reconstructionist idea that God is personal empowerment, strength, and the ability to make a difference. Oprah, who has achieved the status of a multibillion-dollar cultural phenomenon by offering a hopeful spirituality that can be hard to pin down as either religious or secular, often encourages her audiences to "connect yourself to the source, I call it God, you can call it whatever you want to, the force, nature, Allah, the power."[14]

If you believe in Spinoza's god, Dewey's god, Tillich's god, or Oprah's god, we Humanists are your allies and friends. But we believe that calling what you believe in "God" is at best utterly irrelevant to whether you're a good person, and at worst it can confuse and distract others and even you from what is really important.

Atheism and Humanism Today

Today, Richard Dawkins has become arguably the world's most famous spokesperson for atheism. But in his book *The God Delusion,* even he suggests that technically he is only an agnostic. On a scale from one to seven, Dawkins explains, where one is utter certainty that there is a God and seven would be utter certainty that there is no such thing, Dawkins claims only to be a six. But this does not mean that there are no such beings as atheists—or that Christopher Hedges is right to say, "I don't believe in atheists," as in the title of his popular recent book. It simply means that religious and nonreligious people alike are in for a lifetime of unresolved debate and often meaningless discussion if we continue to convince ourselves that what is really important about religion is whether or not we say we believe in God.

There is a better way, and it starts with acknowledging that *atheism,* like *God,* is a complex word. No one has explained this better than Sherwin Wine, who saw atheism as divided into a number of intellectual categories:

> There are different kinds of atheism. The most popular kind is "ontological" atheism, a firm denial that there is any creator or manager of the universe. There is "ethical" atheism, a firm conviction that, even if there is a creator/manager of the world, he does not run things in accordance with the human moral agenda, rewarding the good and punishing the wicked. There is "existential" atheism, a nervy assertion that even if there is a God, he has no authority to be the boss of my life. There is "agnostic" atheism, a cautious denial that claims that God's existence can be neither proved nor disproved; this type of atheist ends up with behavior no different from that of the ontological atheist. There is "ignostic" atheism, another cautious denial, which claims that the word "God" is so confusing that it is meaningless; this belief, again, translates into the same behavior as the ontological atheist. There is "pragmatic" atheism, which regards God as irrelevant to ethical and successful living, and which views all discussions about God as a waste of time.[15]

Wine's point is that most nonreligious people are atheists in one or another of these senses, even if they are pantheists or Reconstructionists or choose to call themselves agnostics or whatever else. But none of these terms has anything to do with what we *do believe*. We must not shy away from atheism as *part of our identity*—part we share with Spinoza, Dewey, Dawkins, Oprah, and millions more. But when *God* can mean anything we want it to, not believing in God is not a very important assertion. What is truly important is to assert, loudly and boldly, that we have a dramatically different way of understanding the world and human values. We are not simply atheists or agnostics or nonreligious—we are Humanists.

If We're Nothing but Selfish Genes, Why Be Good?

A common misunderstanding of the scientific side of Humanism is that we see people as made up of what Richard Dawkins famously called *The Selfish Gene*, and so we must have a vision of entirely selfish people, on their own, pursuing their own ends, on their own time. This is a caricature of Dawkins's insight. The fact that we are biological creatures, made up of genes, does not mean we are totally selfish. Ants are made up of selfish genes too, yet as evolutionary biologists like E. O. Wilson have demonstrated time and again, they are unbelievably selfless, willing without hesitation to die en masse for the sake of their fellow creatures.

In fact, the best science tells us that we humans have the ability to be the greatest cooperators in the history of life on earth; we have evolved the ability to help one another in extremely complex, ongoing, and profound ways. This is in no way to minimize our ability to visit hatred, iniquity, and even genocide upon one another, as both religious and nonreligious humans have demonstrated the ability to do. But clearly we also have the ability to build huge hospitals, create organizations like the Peace Corps, and occasionally have the open-mindedness to work together to elect political leaders who do not look like us or have names that sound like our own. If we can better understand the part of our human nature that facilitates generosity (Hint: it isn't only caused by prayer or church attendance), we might, just might, be able to summon it up more often.

Why are we naturally good as often as we are? Because of something called "the prisoner's dilemma": the idea that people, like prisoners attempting to escape a jail cell, are constantly faced with decisions about whether to cooperate with one another or whether to "defect"—to rat their partners out, to betray them in hopes of maximizing benefits for themselves. This dilemma has long been closely studied in an academic area known as Game Theory, designed at Princeton University before and during WWII, as well as in a more scientific approach called Evolutionary Game Theory, intended to see what Game Theory can teach us about our evolutionary origins.

What we learn from studying the prisoner's dilemma is that sometimes rationality is not so rational—or maybe you prefer to see it as the other way around. In "games" in which you find cooperation, there is always a potential giver, called a "donor," who pays a cost in order to cooperate, and there is always a potential recipient, who gets a benefit from the donor's cost. And so "rational" players—or at least, players with the simplest possible sense of what rationality is—both defect so as not to end up as the donor—and they both get nothing. But more sophisticated players learn that in the long run, giving and getting can occasionally be a win-win strategy. Like teammates on a winning sports club putting aside some individual glory for the sake of a championship, we're all clearly, objectively, demonstrably better off when aiming for cooperation than when aiming for selfishness.

Another way of looking at why we're naturally impelled to be good is what cognitive scientist Steven Pinker calls "a feature of rationality itself," or what you might better understand as a feature of our consciousness. Consciousness is an incredibly mysterious thing, so much so that many theologians and religious thinkers have argued that it must simply have been given to us by God. But that argument is a copout, dodging a much more complex, amazing truth: that primates and many other animals' brains and minds share features with ours, such as the ability to form lasting pair bonds, or recognize themselves in a mirror, or use tools. But our minds are so much more complicated not because God said they should be so, but because, as Carl Sagan described them, our brains have ten to the eleventh power or so of neurons and ten to the fourteenth power of synapses.[16] If you further increased these numbers exponentially, the resulting creature might have as little to say to us as we do to ants.

Thanks to all this neural complexity, we have evolved a complex enough ability to think that we can recognize what Pinker calls "the interchangeability of perspectives"—that if I want you to do something for me, I have to be able to take your interests into account as well, unless I am a "galactic overlord" or am not interested in a particularly high rate of success. Pinker points out that this is the key insight of the golden rule, which a number of different societies and thinkers discovered independently.

The Evolution of Cooperation

Many people assume that human beings are doomed to want to be "galactic overlords"—that the only way you can get us to act decently is to promise eternal reward or threaten eternal punishment.[17] There are even plenty of people who suspect that God, heaven, and hell are not real, but are so intimidated by what they see as the lack of this-worldly incentives to behave ethically that they'll happily go along with incredible God-myths just because such myths at least seem to provide a basis for moral behavior. They don't *believe,* in other words, but they *believe in belief.*

But the "golden rule" is golden because it's a simple, easy-to-understand reminder that there are many reasons to be good, beyond God—and, in fact, God may not even be the real motivating force behind the good behavior of many pious people. After all, we are evolved creatures, and much of our goodness—along with our constant struggle to bring it out—comes from the way we evolved. Evolutionary scientist Martin Nowak has identified five rules[18] to explain why, despite our genes' "selfish" drive to replicate themselves, "Humans are the champions of cooperation: from hunter-gatherer societies to nation-states, cooperation is the decisive organizing principle of human society. No other life form on Earth is engaged in the same complex games of cooperation and defection."[19]

Kin Selection

The first rule of the "evolution of cooperation" is the most obvious. Nowak calls it "kin selection"—as in J. B. S. Haldane's famous comment, "I will jump into the river to save two brothers or eight cousins." It's the mysterious pull we often feel to love, nurture, and come to the aid of members of our own family. From a detached, coldly scientific standpoint, measuring only whose genes get passed on, the lives of our siblings, not to mention our children, mean almost as much to us as our own lives. But it would be a sad world if the passing on of the family genes was the only reason we had to be good to one another. And besides, a wise religious person once said, "Friends are God's way of apologizing for your family"; so this first rule is badly misunderstood if isolated from the other four.

Direct Reciprocity, or Tit for Tat

The next rule is also straightforward. Nowak calls it "direct reciprocity," or "tit for tat." You don't need to be a particularly evolved person—actually you don't even need to *be* human—to feel motivated to help others because they can help you. But scientists watching humans playing games like prisoner's dilemma noticed that we tend toward some interesting variations on "You scratch my back, I'll scratch yours." For example, how do we respond when we've been cooperating successfully with someone for a while, but then they cross us? Is forgiveness evolutionarily sound? What if a trusted teammate or business partner or friend *accidentally* harms us? I once watched a colleague agonize over this when his twenty-something daughter, a little drunk at the time, literally burned his house down by accident. When he found out what had happened, he wanted to hug her forever just because she was alive. And then he wanted to *kill* her. How generous should we in these kinds of situations?

Researchers like Nowak note that the most *efficient* way to play the direct reciprocity game seems to be what they call "win-stay, lose-shift": keep a successful relationship going as-is, and when problems arise, consider a new strategy. But Humanists are not Vulcans. *Efficient* is not always humane or decent. And if we only ever helped close family members or those who helped us, we would be very lonely creatures indeed.

Indirect Reciprocity, or Paying It Forward

Still, why do we tip waitresses we'll never see again, or stop on a busy day to give a passing stranger directions? Is it because we think God will punish us if we stiff her or stiff-arm past him? For some people, perhaps—but ask yourself: is *God* really the first and only thing that comes to *your* mind when deciding to give to a charity? What about your reputation? A bear can help another bear find food, but it can't gossip about how a third bear, a year or two ago, tried to swindle him out of a pot of honey. We humans can and do, and we therefore live our lives with the constant awareness that our behavior may be seen and evaluated by others, for better or for worse. Gossip and reputation may even have played a major role in how our brains got so big and powerful in the first place: perhaps we needed as many neurons and synapses as we have in order to be able to remember a catalog of details about who has been helpful to whom and who has been obnoxious.

We've also evolved the ability to simply "pay it forward": I help you, somebody else will help me. I remember hearing a parable when I was younger, about a father who lifts his young son onto his back to carry him across a flooding river. "When I am older," said the boy to his father, "I will carry you across this river as you now do for me." "No, you won't," said the father stoically. "When you are older you will have your own concerns. All I expect is that one day you will carry your own son across this river as I now do for you." Cultivating this attitude is an important part of Humanism—to realize that life without God can be much more than a series of strict tit-for-tat transactions where you pay me and I pay you back. Learning to pay it forward can add a tremendous sense of meaning and dignity to our lives. Simply put, it feels good to give to others, whether we get back or not.

Network Reciprocity

Churches, synagogues, mosques, and temples are also examples of unselfish human cooperation—and we form them because we evolved to, not simply because we believe in God. From an evolutionary standpoint, a congregation is an example of what scientists call "network reciprocity": clusters of individuals bonding together and making an agreement to help one another without one individual expecting direct return from the next. Such groups, or "networks," tend to be small enough that those who take and take without ever giving back can eventually be rooted out, leaving a situation in which people are accustomed to cooperating with one another and trusting each other. Not surprisingly, Nowak's research suggests that a given population or group of people is most fit and successful when most or all of its members are cooperating in this way.

People need community. Not just out of some whiny desire to be hugged or avoid loneliness—we need community because we succeed best in life when we can count on reliable help from a wide range of individuals with a range of skills and talents, all of whom know us personally enough to treat us as their own when we are in need. You just don't usually get that kind of feeling from working in the same office with people, or living in the same apartment building or even on the same rural street. There are a few exceptions that prove the rule. Sometimes the dorms at a great college can feel like a true community, and then I hear from alumni several years later about how much they miss that experience. Sometimes a tightly knit urban music or art scene can provide a great sense of belonging, but then you usually hear people in these scenes talking about how music is a religion. Even then it's a religion where many people don't trust or help each other as readily as they might in a well-run church. For most people, it takes a congregation. But it doesn't necessarily take *God*.

Group Selection

Another similar rule for why unselfish cooperation evolved is called "group selection": the idea that sometimes individuals may sacrifice their own personal success—even the chance to pass on their own genes—and yet still "win" if members of their group have success against members of other groups. This explains a lot about why human beings seem so universally willing to let big groups define them—we give ourselves up for fellow members of our tribe, race, ethnicity, city, state, or nation. As Darwin said in *The Descent of Man,* there is no doubt that "a tribe including many members who . . . were always ready to give aid to each other and sacrifice themselves for the common good would be victorious over most other tribes; and this would be natural selection."[20]

Of course, if this idea of Darwin's sounds like it's been used to justify one group cooperating to take advantage of another—what's been called "social Darwinism" (though Darwin himself, a staunch abolitionist, would have wanted nothing to do with it)—that's because, unfortunately, it has.

The scary thing about understanding the evolution of cooperation in our families, religious and ethnic groups, congregations, and countries, is that it sounds like an excuse to believe that we should just select the fittest group and everyone else be damned. And so it is crucial to remember that evolutionary theory is only an explanation of what has happened up to now—it is not a recommendation for how human beings ought to behave in the future.

Humanism is not only not equivalent to social Darwinism, it is precisely the rejection of social Darwinism. Humanists recognize that competition between groups has been part of our evolution—there is no way to expunge this basic fact from our minds or our history books—but now that humanity has discovered this, we can and must search fervently for healthy, nonviolent ways for groups of people, as well as individuals, to relate to one another. This also means we do not pretend that evolution is the solution to every problem, or that it is always a sweet, innocent story. When thinking about why we cooperate—and why we are good—we do not shy away from the question: what about the fact that we can be so lousy to one another?

Our job as Humanists is not to minimize the role selfishness and brutality have played in human history. It isn't even to overlook or explain away our own temptation to be cruel. In fact, we need to be honest with ourselves because we have to decide—every day, every minute—which is it going to be? Cruelty has evolutionary value. Kindness does too. But we can't have both at the same time. And not only do we compete and struggle with each other, we do so within ourselves. We have all these competing desires and drives: for food, sexual reproduction, loving acknowledgment, dignity. Humanism is the active choice that, whenever possible, dignity gets priority. It means acknowledging and understanding our selfish genes precisely so that we can continue to evolve beyond them.

If There Is No God, Why Is Belief in God So Universal?

Speaking of evolution—in recent years a popular strategy for making theism (seem to be) more in line with science is the suggestion there is a "God gene." We *must* be hardwired to believe in God, the claim usually runs, or why would such a huge majority of human beings hold such a belief?

And it is true that most people—even a great many who call themselves nonreligious—say they believe in some kind of supernatural God. Why? Was this notion implanted in our brains somewhere, whether by nature or by the deity itself, so that we would go forward faithfully, no matter how many natural disasters, religious wars, and abusive preachers gave us reason to doubt?

When you take a good scientific look at the human mind, God is not a *gene*, but a *spandrel*. A spandrel is the triangular negative space created between two archways when they are positioned side by side, often elegantly decorated in churches and other imposing architectural structures. Evolutionary scientists Stephen Jay Gould and Richard Lewontin first pointed out that this term, and the idea of a spandrel in general, can explain something critical about our God-beliefs: that they are *by-products*, not *adaptations*.[21] Spandrels, Gould and Lewontin argued, are often ornate, beautiful, and seemingly meant to be a focus of attention in an architectural structure. But you don't really *need* to include a spandrel when you're building a cathedral. What are really necessary are the archways, which hold the weight of heavy ceilings better than other forms of support. And when you place arches next to one another, spandrels are just a by-product—the shape that happens to be created by the space between two arches. Well, belief in God is also a by-product—of two of the most important architectural features of our minds: archways of our brains that produce the spandrel of faith—what cognitive scientists call "causal reasoning" and "theory of mind."[22]

Causal reasoning is a self-explanatory term. Our minds evolved to look for the causes of things—not eating causes hunger, watering plants causes them to grow, and so on. If we did not understand that actions have consequences, it would be very difficult to get anything done. When I was a boy I once broke an expensive crystal vase while out shopping with my parents, and offered the feeble excuse "It just happened!" Of course, everyone knew that there was a more direct cause—such as my playing a little too roughly with my brother in a store full of expensive glass. But small children attempt this excuse time and again, because they too have an important insight—that while everything has a cause of some sort, not everything is caused by intentional action. Sometimes nature causes change for no apparent reason—a thunderstorm messes up a perfectly good hair day, a freak injury causes my favorite baseball team to lose the World Series, or gravity causes an ornate vase to fall off a display table and shatter into a thousand pieces (well, two out of three isn't bad).

Whether it's a trivial mishap or a bloody war, we're programmed to believe that everything has a cause, and even when some things don't, we still want to believe they do, so we infer that something we can't see and can't understand caused them. This pattern of causal thinking is so strong, it even outlasts belief in a traditional God. I've known countless people who might check off "none" on a survey of their religious preferences but will still tell you with a straight face that "everything happens for a reason." You might know people who wouldn't be caught dead in the church or synagogue they grew up in, but earnestly consult their horoscope every day or believe that they have a lucky number. We just don't like randomness, so we look everywhere around us for little "signs" that the mysteries in the universe have a purpose and that the strange things that happen to us every day were "caused" by some sort of watchful force.

But again, our fascination with superstition, even those of us who ought to know better, has its own reason. We evolved over millions of years to look for causes, whether they exist or not, because if we hadn't, the world would look even more confusing to us than it does now.

Another way we human beings adapted to our surroundings, creating belief in God as a natural by-product of the structure of our minds, is known as "theory of mind," or "folk psychology," as psychologist Scott Atran calls it. In other words, I can experience myself thinking, and as Descartes recognized, "therefore I am." But all the philosophy or technology in the world is not enough to allow a single one of us to actually get inside another person's mind and experience *that mind* thinking. I can never directly experience "You think, therefore you are"—I just have to assume it. Yet life would be incredibly different if we did not readily assume that just as we have minds, others do too. And once we gained the ability to project a sentient mind like our own onto another *person* whose thought process we couldn't actually experience, what was to stop us from projecting similarly onto animals, plants, mountains, and stars? Despite our ability to reason, it can still feel uncannily logical to imagine that a car, or a tree, or the moon is alive and pulsating with purpose, because we need to have that *confidence* that we're not alone in the world in order to believe that our loved ones are really "in there," thinking just the way we are, loving us the way we love them. We can't get inside their brains to check, any more than we could get inside Zeus's mind if he were really up there, perched atop a thundercloud.

And while anthropomorphic objects may have been a bigger feature of ancient religions than they are of today's, it's not even hard to understand why highly intelligent people can believe so easily in an *invisible* God. After all, when a child's mother leaves the room, he has to be able to learn that she still exists even though she is invisible to him. She'll come back, he internalizes. From there, it's not too much of a stretch to think that God, the being who created our mother and the whole world—is only invisible for now and will return soon too. If Mother can be Mother even while not present physically, how much more so with an allegedly all-powerful Father.

So to those who say that the ubiquity of belief in God is itself a proof of God, we say no. There's much better evidence to support the idea that God is a spandrel, a by-product of the evolution of our minds. Faith in the supernatural is not a mystery. There are reasons for it.

But on the other hand, when atheists who allow themselves to become shocked and indignant not only by religious extremism but even by the fact that some people still believe in God—when we whine that God is like a fairy tale or Santa Claus—at best we are confused, and we may even be irrationally angry at religion. Our minds are what they are—we did not choose for them to evolve this way—and with the spandrel of God-belief so well in place, most people will continue to at least wonder about the idea of some form of supernaturalism, even as we continue to think through—and cease to believe in—more and more different kinds of religious ideas. It makes little sense to say that logic, reason, and science should eliminate or replace all religious beliefs, when that same scientific approach tells us these beliefs have taken millions of years to evolve and are deeply enough ingrained that they will be with many of us in some form for a long time to come.

Still, a decision to embrace Humanism and deny all religious beliefs is liberating. We feel free from having to believe things that make no sense, just so we can think we know the cause of everything. We feel free from having to pretend we know things we don't. We feel free from having to have a relationship with a being we strongly suspect has never existed and cannot be known. We feel free from having to submit to an authority that cannot be questioned and is capricious. And we can even assert, as Margaret Mead did: "Never doubt that a small group of thoughtful, committed citizens can change the world. Indeed, it is the only thing that ever has."

Without God, How Do We Know What Good Is?

There are those who worry that if we try to reduce our understanding of God and morality to a science, we will end up without any criteria for good. This is one of the purest forms of that old question, is it *possible* to be good without God?

There are a significant number of people who believe that we'll descend into moral chaos without a belief in God. This is why so many atheists I know like to joke, when asked what has changed about life since they stopped believing in God, "Well, I didn't suddenly become an ax murderer." And while not many Christians go around looking over their shoulders for an atheist running after them with a bloody hatchet, like Jack Nicholson in *The Shining*, this kind of joke has a kernel of truth to it, because people have always had visceral fears about those whom they don't understand.

In fact, there are many religious people who fear that, although Humanists might not need a God to tell us *how* to be good, we might not be *motivated* to do good unless we felt as if someone was watching us all the time. This was certainly the fear of a family friend of mine who has gone through some rough times with depression and anxiety, and turned to God for help with those problems even though he wasn't really sure he could believe in God. When he found out I was becoming a Humanist, even discussing the subject made him angry and nervous, because—as he admitted—he worried that if my disbelief ever "rubbed off" on him, he might be doomed.

This is the kind of concern voiced by theologians like the Christian prison fellowship leader Chuck Colson, who was incarcerated for his role as a Nixon aide in the Watergate scandal. Colson acknowledges that of course there are many individuals who do just fine and live perfectly good lives without God, but that society as a whole would break down if we didn't have a godly foundation for our public values. So America must remain a Judeo-Christian nation (without much emphasis on the *Judeo*, either), we must put

God on our coins and in our pledge, and we must remind others who are otherwise not as reliable as we are, or as other elites are, that they have a kind of divine security camera watching them.

"But what about the worker bees, the bricklayers?" asked a well-known right-wing intellectual and Ivy League professor when I first met her. She acknowledged a private leaning toward secularism, but seemed to fear that the "common people" were somehow constitutionally incapable of understanding or respecting Humanists and atheists like herself. Ironically, though, no more disastrous an authority on human morality than Hitler also espoused this brand of skepticism about goodness without God:

> This human world of ours would be inconceivable without the practical existence of a religious belief. The great masses of a nation are not composed of philosophers. For the masses of the people, especially faith is absolutely the only basis of a moral outlook on life. The various substitutes that have been offered have not shown any results that might warrant us in thinking that they might usefully replace the existing denominations . . . There may be a few hundreds of thousands of superior men who can live wisely and intelligently without depending on the general standards that prevail in everyday life, but the millions of others cannot do so.[23]

And if this has you thinking, "Aha! I knew Hitler couldn't be all that religious! Maybe we can still blame Nazism on the atheists!" remember that as soon as you get into forgiving bad religious people on the basis that they might not have *really* believed, in fairness you'd probably have to look at many seemingly great people who said they did what they did to honor God but may actually have had much more secular reasons for doing it. Mother Teresa, for example—even if you can accept her as an example of goodness despite her hard work to prevent millions of people from obtaining condoms who desperately needed family planning and protection from AIDS—and I can't—it has lately become well known that she too harbored overwhelming private doubts about God's existence, and indeed may not have believed at all.[24]

A Knockout Punch for Goodness Without God?

Frustrated that the argument against goodness without God is still taken so seriously despite the obvious philosophical holes, I called one of my favorite philosophers, Rebecca Goldstein, a MacArthur Fellow and author of *Betraying Spinoza* and *The Mind-Body Problem*, to talk about it. Right away Rebecca and I agreed that one of the hardest parts of dealing with this issue is just trying to understand the other side. It can be hard to imagine that there really are people who believe we can't be good without God.

Rebecca pointed out that in almost any of the above varieties of concerns, we must deal with what philosophers know as Hume's "is-ought problem." As the great eighteenth-century English philosopher and skeptic David Hume pointed out, there is a huge difference between what "is"—what exists, the way the world is—and the way the world ought to be. One of the basic questions philosophers have occupied themselves with, then, is where do we get our values? Who says something that is one way ought to be another way, if not God?

This is not a mere egghead question that only professional philosophers and theologians deal with. It is the first thought that goes through the head of a young husband and father of three, sitting in an oncologist's office, told that his pancreatic cancer has metastasized and is inoperable. "You may have about six months," the doctor softly informs him. He understands that that's what *is*. But with every fiber of his being he feels: it's not what *ought to be*! And some theologians claim that the only way we can justify believing that it shouldn't be that way is if God told us so.

In *God and the New Atheism: A Critical Response to Dawkins, Harris, and Hitchens*, Catholic theologian John F. Haught asks, "Can you rationally justify your unconditional adherence to timeless values without implicitly invoking the existence of God?"[25]

This question—where do we get timeless values (such as the fact that it is good to heal a sick father of three so that he can live and be with his wife and kids) without God—is a variation of the is-ought problem, and it comes up not just in the doctor's office but in our most important debates about political and social issues. Of course, when it does, we should ask our questioners what they mean by timeless values, inviolable "oughts." They usually cite as an example that murder is wrong, which of course makes you wonder why they aren't pacifists. Or they may say that rape is wrong, which is despicable when you realize they're implying we need religion to figure that one out.

Still, given all we already know about why belief in God is common, and given how wrenching some of these questions about values and justice truly are, Rebecca Goldstein and I agreed that it is at least reasonable that many people would wish for a God who could insist that justice be done. But as for those who would go a step further and suggest that there cannot be any justice or any good without God, Plato's dialogue *Euthyphro,* written in 380 BCE, provides what Rebecca calls the "knockout punch" against them.

In the dialogue, Socrates reminds his friend Euthyphro that a crucial question is not simply whether we can know if one or another particular action is good, but *on what basis* we determine whether any action is good. Euthyphro answers: "Piety, then, is that which is dear to the gods, and impiety is that which is not dear to them."

But Socrates responds: "Is that which the gods love good because they love it, or do they love it because it is good?"[26]

If the former is true, then who says the gods are not evil, unfair, or frivolous? The gods could choose to love anything they want, regardless of whether or not human beings would consider it just. Is that the kind of system we want to live by? Do the gods want us to be blindly, unquestioningly obedient to them even if they behave like murderous scoundrels? And if the gods love the good simply because it is good, then it could damn well be good on its own. We wouldn't need a god or gods to tell us what morality is—we'd be responsible for figuring it out just as they were.

In either case *Euthyphro* drives home the point that mere belief in God can't make us good, and it can't point the way to "timeless values" that we humans aren't equally capable of arriving at on our own terms. Gods don't—can't—create values. Humans can, and so we must do so wisely.

It's not that Christian theologians have never heard of Plato's *Euthyphro,* of course. And it's not that they don't have responses to it that they think work. One example of such a response comes from Paul Chamberlain in his book *Can We Be Good Without God?* Chamberlain argues that goodness was in God's very nature when he created the world, but he left us free to discover that goodness for ourselves. This is supposed to address Plato's concern because it makes a distinction between our not automatically "knowing" that God is good and God's actually being good. But it isn't really an argument at all. It sounds sophisticated, invoking philosophical jargon: "[Those who invoke the *Euthephro* argument] confuse categories of knowing and being." But it assumes that we will agree unconditionally with the book's Christian character Ted when he says, "There really can be no question about the goodness of God's nature." "And why would you say that?" Francine demanded. "Because I see what it has produced." "And what is that?" "The moral values that we've been speaking of all along."[27]

This is not a response to Euthyphro at all. It is merely the statement that goodness is what God says it is, but that God is all good, and we know that from his works. That may sound like a circular argument. And it is. But that is not the worst part of it. The worst part of this argument is that it defies not only common sense but moral common sense. The world is not all good. The moral values we find intrinsic in human beings are not all good. Much of the world we live in is absurd. Ignorant, selfish fools prosper while innocent babies are slaughtered. Football players can pray for touchdowns, but not a single amputee, no matter what the unfair circumstances surrounding her injury, has ever successfully prayed to regrow a limb. There is simply not one single theological argument—whether formulated by progressive theologians like Paul Tillich or Harold Kushner, or in mystical texts like the Kabbalah, or by famous saints like Augustine or Francis of Assisi, or by some other obscure thinker that only "cool" religious people know about—that amounts to a decent response to Plato's *Euthyphro*. Because the fact is, there could be a god that hates amputees.[28] We can neither prove it nor disprove it—ever. But it would be beneath our dignity to worship such a God. It would not deserve our time or our energy. Meanwhile we have much better ways of explaining the world and much better ways of understanding moral and ethical values.

Still, even if we can "knock out" the argument that God is necessary for good, the focus of Humanism is secondarily on God, and primarily on good.

Where Do Our Ethics Come From, If Not from God?

The simplest way to put this is: our ethics come from human needs and interests. What do human beings need to flourish? As the Humanist Manifesto puts it, "Ethical values are derived from human need and interest as tested by experience. Humanists ground values in human welfare shaped by human circumstances, interests, and concerns and extended to the global ecosystem and beyond. We are committed to treating each person as having inherent worth and dignity, and to making informed choices in a context of freedom consonant with responsibility." But where do this inherent worth and dignity come from? It's an important and fair question—one that neither the average religious or nonreligious person has thought a lot about.

Rebecca Goldstein cited her mentor Thomas Nagel, Princeton philosopher and author of the classic *The Possibility of Altruism*, as the best contemporary thinker on this question. Goldstein and others have called Nagel the most important ethicist since Kant.

Essentially Nagel argues that there are certain natural attitudes that already commit us to valuing our own lives—we all know what it is like to feel moral outrage when we are wronged, insulted, beaten, discriminated against. Nagel says that logic commits us to universalize from here—to reason that we all know for ourselves that there is a right and a wrong, and we know that we cannot exist alone, so from there only radical selfishness could prevent us from understanding that these concepts are universal—and radical selfishness leads to great unhappiness, so that's not an option either.

As Rebecca says, that is the basic intuition of Kant, and of all ethics—"ethics really isn't all that complicated."

Still, the idea that our entire concept of goodness is based on human needs can take a little getting used to. When I was first learning about this aspect of Humanism, it seemed a bit offensive to the part of me that had been taught, by the Holocaust survivors in my Jewish community, that we should never again make the mistake of treating any group as inherently inferior—whether the group was Jews, blacks, gays, Palestinians, the Sudanese, or anyone else. What about nonhumans, I wondered? Does Humanism, with its semantic focus on humans, leave a door open for the mistreatment of animals? Of nature in general? Is it . . . anthropocentric? Or even "speciesist"?

But no matter how much we may value animals such as monkeys or dogs—and we should indeed concern ourselves with behaving ethically toward them, lest we lose our own humanity in treating them with feckless-ness or disdain—I had to admit it would be absurd to ask the question: "Is it *ethical* for monkeys to scratch themselves in public?" I realized it would also sound a bit strange to suggest that it was loose morals for dogs to go around having sex with so many bitches.

Likewise, it is incredibly important for human beings to deeply value the whole environment, from the oceans to the forests to the atmosphere. We have no right to steal the clean air and water and the diversity of plant and animal life that we know future generations will desperately need. No one appointed us Kings of the Universe. But we also don't ask questions about whether the oceans are behaving ethically toward *us*. We could rightly be accused of madness if we were to accuse the rain clouds of attempting to drown us or the ozone layer of not being sufficiently tolerant of our skin's needs. (Though of course these are very close to the kinds of value judg-ments we used to make in older times, when we had no scientific reason to doubt that the natural world was as animate as we are.)

No, values and ethical behavior, until such time as other similarly sentient beings make themselves known to us, can only really be found in human beings. Our focus on human beings in choosing the term *Humanism* is nothing more nor less than the acknowledgment of this fact.

Keepers of the Question

Our morality is based on human needs and social contracts, and these things are not perfectly, eternally objective. After all, slavery was once considered morally acceptable by almost all religious people, including Christians. If values were timeless and objective, either the early Christian saints who believed in it were horribly wrong, or values change.

But the point is that some social contracts are much better than oth-ers. Not only do we not need "objective" values to condemn heinous crimes and uphold ethical standards, we cannot ever be confident that objective values exist. We can postulate them, but there is no way to prove them right or wrong, existent or nonexistent. What proof would suffice? You'd have to have divine revelation—in which case, if it comes, we Humanists are perfectly willing to change our minds. But we're not holding our breath.

Meanwhile, there will always be competing systems of proposed "objective" values, meaning that we will be at the mercy of their earthly representatives. If you sign on for the idea that we need objective values, you are signing on for a lifetime of placing great importance on the often petty bickering among ministers, priests, rabbis, imams, swamis, and other gurus, as to which one of them possesses the *truly objective* values and why *all the others* possess only false words and ill-begotten human inventions. Is that what we want? To submit our psyche—the only brain, the only intelligence we will ever possess—to the mercy of one holy man after another parading various subjective arguments for why we should obey them on the basis of their supposedly objective truth? If this were the only way, maybe we would. If there were no other way to feel that life was worthwhile or that we could be part of a community, maybe we'd sign on for all this despite its flaws. But there is a better way. Humanism (or whatever other word you prefer to use for goodness without God) is that way.

Besides, if there are objective values, then anything can be justified in their name. If no values are eternal, then no matter who tells us what to do, we must always question. No order to murder can be blindly obeyed, blithely excused: "I was just doing what I was told." Who told you? Why listen to them without also applying your own reasoning, your own heart? If no morality is absolute, no war can be justified by the fact that the one true God is on our side—and this notion of absolute morality favoring one side of a conflict has been used to justify almost every war ever fought. Will people then be less motivated to fight? Yes, most likely—and isn't that a good thing? Shouldn't we be slower and more hesitant to ship our sons, and now our daughters too, into bloody battle? If there are really good reasons for defending ourselves, reasons beyond the idea that God says so, then people will rally behind them. God is not necessary for such things and can be a hindrance to clear thinking.

And if no morality is timeless and eternal, then we will never be able to fool ourselves into thinking that there is one set of easy and obvious answers to questions about euthanasia, abortion, capital punishment, or other such issues. We'll have to argue them out, with neither conservatives *nor* liberals ever able to say they are right in every case, without thought. What is so wrong with this? Indeed, we Humanists can take pride in our passionate belief in a morality based on unfettered inquiry, on compassionate questioning. Call us "the keepers of the question." We are proud to welcome a future of permanent debate and discussion about moral issues, a world in which we will never stop refining our views, never stop exploring how we can promote human dignity more effectively, never stop trying to better understand and more effectively eliminate human suffering.

And are there religions that understand the need for debate, discussion, and critical thinking? That can't be painted with this cynical brush? Of course, most progressive churches and synagogues in this country don't fall prey, at least not entirely, to magical or fundamentalist thinking. But they also don't have any more "objective values" than we Humanists do. Not Unitarian Universalist or United Church of Christ churches. Not most liberal Presbyterians, Episcopalians, or Methodists. Not Reform, Conservative, or Reconstructionist Jews. And not progressive or liberally religious Muslims either. Such folks wouldn't be caught dead trying to persuade you that it's their way or the highway. They are to be admired for this—and they are the allies of Humanism. Therefore, if you subscribe to one of these religions, you are no more involved in the enterprise of "objective values" than I am. You are a living example of the fact that subjective values can be wonderful values.

CHAPTER TWO

A Brief History of Goodness Without God, or a Short Campus Tour of the University of Humanism

The vision of an ethical life that I have been describing is an unabashedly new phenomenon. Humanists—thoughtful, positive-minded atheists, agnostics, and the nontraditionally religious—hold many values today that are simply different from the values of our ancestors. Yet for me this feels like a mischievous, indeed slightly blasphemous thing to say. We may openly crave the latest and most advanced updated computers, cars, or iPods. But when it comes to morals and values?

We like to imagine ourselves securely embraced by the past—as reading from the same religious and ethical playbook as our grandmother's grandmother's grandmother. It doesn't feel particularly inspiring to imagine that that great-great-great-great-great-grandmother would most likely have stared at us in open horror for calling an ambulance and attempting CPR on a heart attack victim before praying for God to banish the demons inside him. It's not heartwarming to be reminded that she would most likely have been disgusted by our support for gay marriage, and taken personal offense at our willingness to vote for a political candidate of a different race or ethnicity.

Discussing religion stirs up our innate loyalty and fear of the novel (our predilection for new gadgets is the exception, not the rule), often touching the conservative side of even the most liberal among us.

This is why a number of liberal and New Age religious movements are built upon the insights of modern Humanism discussed in the last chapter, but dress them up in old clothes to create an aura of exoticism, or tradition, or better yet a mix of both. In fact, millions of people who say they believe in God also live by Humanistic ideals but, craving an emotional connection to the past, allow themselves to be convinced that modern morality is actually cut, whole cloth, from the ancient religion of their ancestors.

But Humanism has a proud past as well. And we all should be more aware of it, even the most religious people. Just as Christians would do well to study Christianity and Islam, and Muslims vice versa, all of us—religious and nonreligious alike—will benefit from better understanding that there is a proud tradition of seeking goodness and wisdom without a God. A number of different historical settings show how and where the key beliefs and motifs of modern Humanism emerged. This isn't a comprehensive history. The history of atheism, doubt, and skepticism have been explored in a number of excellent books of late, and most of those spend at least some time exploring the positive, life-affirming aspects of atheism and agnosticism to which we might refer as Humanism. Jennifer Michael Hecht's brilliant *Doubt: A History* stands out as a good place to begin.

I also won't spend any time here on figures such as Sir Isaac Newton, who even given his Christianity could be considered a precursor to modern atheism and free thought because his discoveries helped lay the groundwork for some current beliefs. Such nuances are much better explored in longer histories like Hecht's. Suffice it to say that the history of Humanism is a broad history, in some way touching nearly every great human idea along its way. For example, the Biblical scholar Richard Elliot Friedman theorizes that God's character is pulled back through the course of the Old Testament—at first he shapes the world actively, literally sculpting Adam out of clay and blowing on him to bring him to life. But then he recedes, becoming less and less involved as the story goes on, and this may well be the result of the Bible's having been written over a long time by writers whose beliefs grew less anthropomorphic as the generations passed. Thus the old Yiddish saying, "It used to be that angels walked on earth, and now they're not even in heaven."[1] Here we find Humanism peeking out from behind the rows of history's cornstalks, inviting us into the field to find it in some of our ancestors' imaginations if not in their explicit words and deeds.

On the other hand, if I were to discuss only those who have actively called themselves Humanists in the same sense that I do today, we would have a very short period of time to study—perhaps a century or so—though we would be able to consider many of the most famous and respected intellectuals of that century. I am reminded of an excellent class I took on existentialist philosophy as a young graduate student, where my professor, R. Lanier Anderson, was quick to point out that only two philosophers of any historical significance—Jean-Paul Sartre and his lover Simone de Beauvoir—actually called themselves "existentialists." Even Sartre's great contemporary and kindred spirit Albert Camus, considered an existentialist by nearly everyone with an interest in the subject, never actually labeled himself as such. So Lanier took us on a wide-ranging tour of the moments and figures in the history of philosophy from Rousseau to Kierkegaard to Nietzsche and more, who, he argued, should be considered either the roots of existentialism or its fellow travelers. We were the better for this kind of generous approach, learning by way of such context to appreciate the heart of intellectual history rather than obsessing over labels and minor details.

Here too my interest is not in who called whom what—atheist, agnostic, Humanist, doubter, infidel, skeptic, deist, or even believer—but who, in the essence of his or her life and work, planted the seeds and paved the way for our ability, today, to be good without God.

This is a kind of abbreviated campus tour of the University of Humanism, stopping at some of the most important landmarks and monuments to great moments past, listing a few famous alumni whose names you might recognize.

The First Atheist, or Ancient Indian Humanism

There are certain mysteries that will remain mysteries forever. What did come first—the chicken or the egg? And who was the first atheist? We can never know. But if I had to guess, I'd say that he or she (the atheist, not the chicken) was among the very first handful of people. You can picture what most likely happened: the first time someone came up with a theory about God, or gods, or goddesses, one of his or her family members scowled, bushy eyebrow raised, and grunted the equivalent of "Don't be ridiculous!" If religion is ancient, then Humanism and atheism are most likely almost as old, because as long as we humans have believed, we have also doubted.

The *Rig Veda*, a thirty-five-hundred-year-old Sanskrit religious hymn, contains the following lines: "Who really knows? Whence is this creation? The gods came afterwards, with the creation of this universe. Who then knows whence it has arisen? Perhaps the universe formed itself. Perhaps not—the one who looks down on it, in the highest heaven, only he knows. Or perhaps he does not know."[2]

Here we have an Indian thinker over half a millennium before the early parts of the Bible were most likely written, speaking brave doubts that would give the pope pause even today. And several hundred years later, in South Asia, such doubts began to crystallize into probably the world's first completely atheistic and Humanistic schools of thought: the Lokayata and Carvaka, two very like-minded groups of philosophers from the early and middle of the first millennium BCE, respectively.

The Lokayata and Carvaka (the origins of their Sanskrit names have been much discussed but are ultimately obscure) have been noted by historians of Indian philosophy for their relentless dedication to truth and critical thinking. But it is the tremendous intellectual and social courage that most stands out in their philosophy, preserved in the writings of the enemies and rivals they riled among almost every one of the many traditionally theistic schools of thought in ancient India.

Three millennia ago, the Lokayata and Carvaka looked out at the world around them with sober eyes and recognized what you and I will see today if we will only steel ourselves to reality: no one has ever been able to prove that he or she has witnessed a miracle. No man or woman has ever risen from the dead. No god has ever appeared on earth to explain how or why he created the basic elements that seem upon any careful, serious examination of the facts to be all we have and all we are. And those who beg and berate us to *believe*—that is, to believe the *unbelievable*—almost always have their own self-serving agenda.

Thus these ancient Humanists, from a time long before that word (and long before the word *Hindu* too, though we commonly look back and retroactively call their religious contemporaries Hindus) systematically deconstructed the main elements of the supernatural religions they found around them and instead taught goodness and righteousness for the sake of *this world.* There is no life after death, they proclaimed, so offer kindness to all, not in the next life but now, today:

If he who departs from the body goes to another world,
Why does he not come back again,
Restless for love of his kinfolk
It is only as a means of livelihood
That Brahmins have established here
Abundant ceremonies for the dead—
There is no other fruit anywhere.
Hence for kindness to the mass of living beings
We must fly for refuge to the doctrine of Carvaka.[3]

But the Carvaka and Lokayata agenda extended far beyond mere amorphous "kindness." They spoke out passionately against the practice of sacrificing elaborate dishes and valuable delicacies to the gods while people around them starved:

If a beast slain as an offering to the dead
Will itself go to heaven,
Why does the sacrificer not straightaway offer his father?
. . . If our offering sacrifices here gratifies beings in heaven
Why not make food offerings down below
To gratify those standing on housetops?[4]

The Brihaspati Sutra, a central Carvaka document lost to the ravages of time but preserved by commentators scandalized by its prescient gospel of justice, speaks out against not only the Hindu priesthood but also the caste system, in ways that remained politically difficult well into twentieth-century India. Even Gandhi, who at times succumbed to pressure to rally the support of rich Indian Brahmins at the expense of the downtrodden Untouchables, was hard-pressed to echo the Carvaka message, "nor do the actions of the four castes, orders, or priesthoods produce any real effect."[5] It is enough to make us guess that there must have been truth in stories like the one in the great Indic epic *The Mahabharata*, of a Carvaka executed on royal orders for harshly criticizing a militaristic ruler.

Sadly, all that remains of the Carvakas in the memory of most contemporary Indians is the misunderstanding that they stood for "eating, drinking and being merry." The Carvaka name is today associated with rich, oily foods that symbolize the good life in India (a misconception rooted in the Carvaka belief that valuable, nutritious ritual meals should feed ordinary people, not gods or temple priests). It's ironic and probably not coincidental that most Westerners have a similar and equally ignorant impression of the most Humanistic school of thought from the ancient West—the Epicureans, who emerged in ancient Greece a few hundred years after the Lokayata in India.

The Epicureans

Three centuries before Christ lived to have his divinity doubted, the Greek philosopher Epicurus said:

> *Nothing to fear in God;*
> *Nothing to feel in Death;*
> *Good can be attained;*
> *Evil can be endured.*[6]

In many ways, this still works today as a Humanistic credo. The point of Humanism is not whether or not a God exists, but whether we ought to worship, fear, or pray to it. Epicurus, along with the school of philosophers he inspired, including the great first-century BCE Latin poet Lucretius, believed the world was composed entirely of atoms that operated according to natural laws. The world might have been created long ago by distant deities of some sort, but Epicurus saw no evidence that these gods cared for humans, or that they were in any way relevant to our lives. They did not answer prayers. Worshipping them said much more about us than about them. If worship was an attempt to live well, according to good values, Epicurus did not despise it, but he affirmed that there was a better way to live. He demanded and sought to inspire courage—there is nothing to cower from in life or death, he said, "for life has no terrors for him who has thoroughly understood that there are no terrors for him in ceasing to live."[7]

As Jennifer Michael Hecht explains, Epicurus's worldview was not a religion exactly, but it went beyond mere doubt in God or abstract philosophizing about the world, instead affirming what Hecht calls a "graceful life philosophy" that recommended practices for improving the human experience: "Those things which without ceasing I have declared unto thee, do them, and exercise thyself in them, holding them to be the elements of right life."[8]

What did Epicurus declare we should do? He insisted we must examine all that we do, all that we choose to love and value, and choose only that which is worth choosing—that which produces what is often translated into English as "happiness." But to say that Epicurus championed human *happiness*, if we take only the simplest and most commonsense meaning of that word as shallow, fleeting pleasure, is to cheat the author of this ancient Humanistic wisdom. And it is to cheat ourselves. Epicurus believed in the value of *pleasure*, but not just *any* pleasure. He was devoted to that quality we might simply call human dignity. He wrote:

> When we say, then, that pleasure is the end and the aim, we do not mean the pleasures of the prodigal or the pleasures of sensuality, as we are understood to do by some through ignorance, prejudice, or willful misrepresentation. By pleasure we mean the absence of pain in the body and of trouble in the soul. It is not an unbroken succession of drinking bouts and of revelry, not sexual lust, not the enjoyment of the fish and other delicacies of a luxurious table, which produce a pleasant life; it is sober reasoning, searching out the grounds of every choice and avoidance, and banishing those beliefs through which the greatest tumults take possession of the soul. Of all this the beginning and the greatest good is wisdom. Therefore wisdom is a more precious thing even than philosophy; from it spring all the other virtues, for it teaches that we cannot

live pleasantly without living wisely, honorably, and justly; nor live wisely, honorably, and justly without living pleasantly.[9]

Also in India, ancient Jains demanded philosophical and theological truth in the starkest terms: "If God created the world, where was he before creation? If you say he was transcendent then, and needed no support, where is he now? No single being had the skill to make this world—for how can an immaterial god create that which is material?"[10] We have numerous additional examples of religious skepticism in ancient Greco-Roman thought—indeed there is the Skeptic school itself, much of it affirming, alongside Protagoras, that "man is the measure of all things." And much of ancient Far Eastern thought is deeply concerned with human goodness without placing much if any stock in the importance of gods or spirits—as when Lao-tzu pointed out that "If lightning is the anger of the gods, the gods are concerned mostly with trees."[11]

But it should not seem strange that the two greatest examples of ancient Humanism, the Epicureans and the Carvakas, are both remembered by most people today as if all they had stood for was good food. Epicurus's name is rarely invoked outside an obscure graduate philosophy seminar unless it is in the name of a restaurant or a recipe Web site like epicurious.com—as if these philosophers amounted to little more than an ancient episode of *Iron Chef.* Both schools of thought can be much more easily dismissed that way. Those theists who feel threatened by this sort of spare, sleek, dignified message of godless goodness have always dismissed Humanism with the wave of a hand, chortling that not worshipping God means little more than worshipping one's own stomach. But our best response is living in the spirit of these ancient exhortations to wisdom, courage, justice, and dignity. We can affirm their memory by appreciating that we are the children, the descendants, of a great but misunderstood wisdom with its roots in the ancient East and West—deep in the soil of the human spirit.

The Imperial Monotheisms

Of course, our roots did shrink back a little farther underground for over a thousand years, at least in the West, between the rise of Christianity as the official religion of the Holy Roman Empire in the early fourth century

CE, until the Renaissance and the Enlightenment. Humanism and atheism went nearly underground for so long because the new Christian empire, and the new Muslim empire that joined it from the seventh century onward, each devoted themselves to crossing Europe and the Near East promoting (to put it mildly) their competing monotheistic dogmas. In such an environment of tension and frequent all-out war over the nature and minutest preferences of the one and only God, it is not hard to understand why it might have been a bit dangerous to suggest that not only was one or another of these dogmas incorrect, but perhaps there was no God at all to be worshipped.

Still, there are a number of holes in the commonly held view that the West was nothing but an intellectual wasteland of narrow-minded religious obedience in the so-called Middle and Dark Ages. Though it may be persecuted and mistrusted, goodness without God never vanishes entirely. Human beings will always strive to discover and live the good life, and there will always be those who challenge the established powers over their prescriptions for how everyone ought to pursue it. Jennifer Michael Hecht beautifully describes the history of Humanism during this period as a "loop-the-loop" of rationalism traversing the Mediterranean region—where "each time doubt traveled on, it was being stamped out in the last place it left."[12]

Some in the Muslim world demonize Humanism and secularism today, and in turn there are secularists and Humanists who describe Islam as intrinsically worse or more backward than other religious traditions. But it was Islamic culture that produced some of the earliest and most profound proto-Humanists of the Middle Ages.

And in fact, given Islam's (only partially justified) reputation for being the most closed to modern thinking of any of today's religious traditions, note the strikingly powerful language in which some of the early Muslim world's atheists chose to express themselves—like Ibn Al-Rawandi, an Arab who is said to have written the following, addressed to God, in the ninth century:

Thou didst apportion the means of livelihood to Thy
Creatures like a drunkard who shows himself churlish.
Had a man made such a division, we should have said to him,
You have swindled. Let this teach you a lesson.[13]

But the world of Islam in late antiquity produced kindred spirits of modern Humanism who espoused much more than just this sort of antitheistic sentiment—in fact, it produced men of the caliber of Abu Bakr al-Razi, called "the most creative genius of medieval medicine," a truly great philosopher, chemist, doctor, and humanitarian, who was both deeply engaged with his Muslim and Persian heritage and "the greatest 'agnostic' of the Middle Ages, European or Oriental."[14] Though he positively affected many lives in this world, he was no less a skeptic about eternal life than Epicurus or the Carvakas. "I know not whither I shall roam," he wrote, about the day when his physical existence would come to an end.[15]

Meanwhile, al-Razi was determined to use this life to carefully study all the religions of his time, mining them for whatever worldly wisdom he could find while indulging the truth claims of none: "Jesus claimed that he is the son of God, while Moses claimed that He had no son, and Muhammad claimed that he [Jesus] was created like the rest of humanity . . . Mani and Zoroaster contradicted Moses, Jesus, and Muhammad regarding the Eternal One, the coming into being of the world, and the reasons for the [existence] of good and evil."[16]

Al-Razi's wide-ranging intellect was complimented by quite a number of Arab and Persian intellectual contemporaries who were greatly respected in their time and in some cases influential far beyond it, despite a deep and biting skepticism about the supernatural aspects of Islam—for example, Abu'l Walid ibn Ahmad ibn Rushd, or Averroes, as his name was Latinized, who lived in twelfth-century Spain and North Africa. Without ibn Rushd's work in translating and commenting on Aristotle, the latter might scarcely have made it from ancient Greece to the modern world as a subject of serious study. This portion of the story of goodness without God could go on and on. The point is simply that those of Arab or Muslim heritage today who are considering the merits of Humanism and nontheism need not feel that they are turning away from their own heritage, because Humanism is as much theirs as it is anyone's. Humanism belongs to all of us.

The Trial of Spinoza as Historical Way Station

As much as the contemporary Humanist or positive atheist can draw inspiration from these episodes in history, finding kindred spirits in great ancient and medieval minds, there is a difference between their ideas and contemporary Humanism's—because contemporary Humanism places greater importance on scientific knowledge that simply was not available in earlier times. Baruch (Benedict) Spinoza symbolized the often painful transition from antiquity to the early days of modernity and modern Humanism.

Born in 1632, Spinoza was a brilliant young man who, at the age of twenty-three, was exiled for life from the Amsterdam Jewish community into which he was born. Spinoza was banished for a heresy—and although we have no records of exactly what Spinoza said at the time of his expulsion, his later writing reveals a man whose ideas were indeed revolutionary for his time, though today he would be considered quite in the mainstream for his beliefs that the Bible was written by multiple authors; that there is no such thing as a single "chosen" people; that reason is more important than faith and tradition; and that God is the universe itself—all-encompassing but unfeeling, utterly impersonal, and unconnected to human affairs.

In his play *New Jerusalem: The Interrogation of Baruch de Spinoza at Talmud Torah Congregation: Amsterdam, July 27, 1656*, David Ives depicts Spinoza in his home synagogue, congregation Talmud Torah, as a charismatic, passionate young man, if a little nerdy—a bit of a *nebbish*—confident that he can defend his new definition of God against charges of atheism. But he is finally unable to restrain himself when his rabbi challenges him to affirm the medieval Jewish philosopher Maimonedes' Articles of Faith—thirteen statements of medieval Jewish theological doctrine that still form the backbone of modern Orthodox Jewish theology today. Spinoza was prepared to argue that nature should be called God, and that since he believed in nature, he was therefore not an atheist. But for this precocious and principled thinker, the idea of a God beyond nature itself, performing miracles or resurrecting the dead, was a travesty—a heresy in itself. And so (at least as Ives imagines the fateful moment of Spinoza's conviction for heresy) the young man lashes back, "No!"—barely able to contain his disdain—when his childhood rabbi asks him to affirm Maimonedes' "I believe with perfect faith that Moses is unsurpassed by any prophet," and "I believe with perfect faith in the resurrection of the dead." And for this refusal he is banished for life from the people he loves, condemned to die of tuberculosis caused by the lens grinding he would do to support himself for the rest of his short life.

Baruch Spinoza was not the first heretic, Jewish or Christian, to be banned from a community. He was not the first to suggest a nonstandard definition of God, or the last to be exiled as an atheist. He was not even the first to be censured for using radical new philosophical and scientific ideas such as those of Descartes, Copernicus, and Galileo to challenge religious truth. In fact, Spinoza was aware that only decades before him, a Christian, Giordano Bruno, and a Jew, Uriel da Costa, had each taken outspoken interest in the religious implications of these new ideas and had died violently because of it. But Spinoza pioneered a new path that was unheard-of in his time and yet is emerging as more prominent today.

Spinoza did not ask to be readmitted into traditional Jewish society after his banishment for apostasy. And he did not accept Christianity, despite many entreaties from colleagues, students, and other Amsterdam Christians. Nor did he become a member of any other religion. Rather, Spinoza lived the final two decades of his life outside of religious identity and community entirely, contemplating his own vision of what it meant to be human. And though he explained that vision to others through writing and teaching, he possessed no cultish desire to convert them. He was arguably the first public Humanist in modern Western history.

Enlightenment, Democracy, and the Pursuit of Happiness

In the century following Spinoza's death, an influential movement called the Enlightenment took shape in Europe and America. This was a loosely formalized but intellectually coherent drive for reform in every area of human endeavor—pushing toward democracy, against not only monarchy but religious control of nation-states, and toward dramatic expansion of the role in public and private life of science, critical thinking, philosophy, and the arts. The Enlightenment, which we might also call a "secular revolution," spans much of the three centuries leading up to our own, a period in which the entire concept of what is "good" changed irrevocably—especially what we consider to be good knowledge, good ethics, and good politics.

Before the Enlightenment's secular revolution, and before the scientific and cosmological insights of Copernicus, Galileo, and Newton—all believing Christians who also provided intellectual building blocks for today's Humanism—"good" knowledge meant knowing what the Bible or another sacred text said you should know, usually as interpreted by a powerful, elite class of priests. Now, in the wake of the Enlightenment, even the vast majority of religious people acknowledged that the best way of knowing anything about the world around us was the scientific approach, accessible by means of education to any person: careful observation, testing, and reliance on actual evidence. And with the new science producing so much damning evidence against any literal reading of the sacred scriptures of the time, it was natural that a new understanding of God came to dominate the Enlightenment—that of *deism*. Deism was the notion that there may have been a God who created the universe, but that this God did not appear to interact with the world other than by assigning nature's laws; therefore, miracles such as the virgin birth, resurrection, or the trinity were impossible. (Those considering themselves deists rather than atheists or agnostics today could also reasonably call themselves Humanists, provided they did not feel the need to worship the creator God or look to it for supernatural instruction on how to live a good life.)

The whole concept of what a good life ought to look like also changed with the secular revolution. A good life used to mean a life of suffering. Why? Because nearly everyone was suffering so much from lack of decent food, shelter, medicine, and leisure time that the best way to prevent panic was to assert that "your suffering is *good* for you." So Jesus became a suffering role model. Buddhism cultivated meditative techniques as an escape from worldly suffering. And African American slaves sang of the redemption their protracted pain would bring them in the next life.

But the Enlightenment propagated a new (to most people), Humanistic view of a good life. This new view was made possible by new science and technology that made commerce, communication, and existence in general easier. It was motivated by horror at the centuries of religiously inspired mass murder that had terrorized Europe. It was influenced by Epicurus and Lucretius as well as the Roman Cicero and other early human-centered thinkers. And it was expressed in manifold ways by brilliant writers such as Voltaire, Rousseau, Kant, Hume, Jeremy Bentham, and others whose work is now considered among the foundation stones of contemporary Humanist philosophy. But our view was first "canonized" in the Declaration of Independence, most likely by Thomas Jefferson: that all people are equally deserving of an opportunity to pursue happiness and to be free of suffering in this life (rather than redeemed by it in the next life). My late mentor Sherwin Wine used to say that he knew his mother had a pre–secular revolution mind-set because she didn't understand how to be "happy." Suffering, she could take. But happiness? Oy! What is there to be so happy about—the world is a mess!

The perception of just what a "good" political system is also evolved during the Enlightenment. In ancient times, most people expected political oppression. There was a vague sense that we could do better. There were even outspoken prophets who vigorously called for justice. But even in the broadest-minded corners of Athens there was never reason to expect, say, a universal, representative democracy. Today, every leader and every society in the world have to address the idea of democracy for all: it is the benchmark of good government. Of course, not everyone has it; sometimes it seems that not even we Americans do. But we expect it, and we know that political freedom and self-determination are basic components of a good life. This view was pioneered by American leaders like George Washington, James Madison, Thomas Paine, and others—and the secularism and Humanistic tendencies of those great patriots is well chronicled in Susan Jacoby's classic *Freethinkers: A History of American Secularism.*

The story of Thomas Jefferson, perhaps the world's first truly Humanist head of state, is particularly worth recalling here. Jefferson was known as the Virginia Voltaire, but he had more in common with another Humanist. "I too am an Epicurean,"[17] he wrote to a friend, about his hope that "the human mind will some day get back to the freedom it enjoyed two thousand years ago. This country, which has given the world an example of physical liberty, owes to it that of moral emancipation also."[18] Jefferson's exact beliefs have been much debated, because he did occasionally speak of a belief in God. But it is clear that Jefferson's was the god of deism—the "Almighty God" who, in the words of the preamble to his 1779 Virginia Statute on Religious Freedom, "created the mind free" and interfered in human affairs for neither good nor ill (and most certainly not to resurrect Jesus from the grave). That 1779 statute gave our country and ultimately the world a wonderful model of true secular democracy, stipulating that Virginia was not a Christian commonwealth and ensuring that, unlike most other states at the time, Virginians would neither support any state church with their taxes nor be discriminated against for lack of membership in it, or in any church or house of worship.

The presidential election of 1800 pitted Jefferson against the incumbent, John Adams, in the first time that religion was ever a major issue in a presidential campaign—and the first case of a presidential candidate being swiftboated for his religion.

Alexander Hamilton, playing a Karl Rove–like role in Adams's campaign, ran this ad for the Federalists:

THE GRAND QUESTION STATED

At the present solemn and momentous epoch, the only question to be asked by every American, laying his hand on his heart, is

"Shall I continue in allegiance to

GOD—AND A RELIGIOUS PRESIDENT;

Or impiously declare for

Jefferson—and no god!!!"

Jefferson turned the tables, portraying his opponents as reactionary Presbyterians. Many religious groups of that era feared that conservative Presbyterians (the 1800 equivalent of today's Christian Right) sought to roll back or reverse religious freedom. Reflecting on defeat, Adams wrote to Mercy Warren years later, "With the Baptists, Quakers, Methodists, and Moravians, as well as the Dutch and German Lutherans and Calvinists, it had an immense effect, and turned them in such numbers as decided the election. They said, let us have an Atheist or Deist or any thing rather than an establishment of Presbyterianism."[19]

As a statesman, Jefferson promoted and defended religious freedom, helping to build for our nation, as an example to the world, what he called a "Wall of Separation" between church and state. He designed the University of Virginia as the nation's first truly secular institution of higher learning—a temple to knowledge and human reason. And later, in retirement, he rewrote the New Testament, cutting and pasting its pages to remove all contradictions or miracles, leaving only stories of Jesus as a human philosopher he greatly admired. *The Life and Morals of Jesus of Nazareth*, or the "Jefferson Bible," remains an inspiring symbol of Humanism today—here was Jefferson, the great leader and champion of democracy, honoring his religious and cultural heritage while refusing to accept it at face value without applying to it his own critical intellect and all the wisdom he could muster.

The Enlightenment was a key moment in the history of Humanism. For the first time, many ideas were not only the intuitions of a passionate but tiny minority, but they rose to prominence, even helping to shape the way whole societies were organized. However, as the previous examples from the ancient and medieval world should demonstrate, Humanism did not begin during the Enlightenment and is not reducible to it. The eighteenth century had its own particular concerns, and even its most brilliant thinkers made their mistakes—Jefferson's more than passing dalliance with slavery being just one. Calls for building a modern way of life based on the "principles of the Enlightenment" can be stirring and have the best of intentions, but often make the task ahead seem a bit too simple, too easy. Modern secular life has brought new challenges of its own—not all of which can be blamed on religion and tradition—and for those we will need entirely new solutions. Still, we will also need strength, courage, and wisdom, and as God is not available as a source of such blessings, we will need to find them in ourselves and each other, and surely we will look to our early precursors in Enlightenment Humanism as significant sources of inspiration.

The New Atheist Four Horsemen of the Apocalypse: Freud, Marx, Nietzsche, and Darwin

Today we are on the brink of a time when millions of people, dissatisfied with the options traditional religions offer them in a world of radical new challenges, yet unwilling to surrender to apathy or cynicism, rally to the idea of goodness without a God. This stark, confident movement of positive atheism and agnosticism we are now building could not have come about without a sharp break from earlier eras when belief in God was required for respectability and social standing. Even Jefferson and other secular, skeptical Enlightenment intellectuals needed to give lip service to belief in a God. Their deity may have been an impotent, nonsupernatural metaphor, but it was also a sign that public atheism was not yet a safe option.

Some would argue that the handful of early twenty-first-century "new atheist" writers are the ones that have accomplished this break from the religious past, and indeed their accomplishments have been impressive. But the true separation from any remnant of premodernity was achieved by the original "new atheists"—men like Charles Darwin, Karl Marx, Friedrich Nietzsche, and Sigmund Freud—and we are still sorting through the ramifications of their work today. Like today's new atheists, the members of the nineteenth century's foursome were far from the only ones doing what they were doing or thinking what they were thinking. But they were enormously influential, and their work touched almost every area of intellectual and moral life.

Darwin's achievement—revolutionizing our understanding of the origins and development of all life on earth by explaining the process called evolution—can be appreciated by theist and atheist alike. Though we Humanists review the data that evolution presents and determine that naturalism and nontheism are the only workable theological conclusions we can come to about the world Darwin's theory describes, we recognize that one can disagree with intelligence and honor, as have even some great scientists such as the geneticist Francis Collins, who is a serious Christian. Certainly Darwin's life and thought have been so extensively documented and analyzed that a rehash here is far from necessary. Still, he was beyond doubt one of history's greatest contributors both to goodness and to godlessness.

On his youthful voyage to South America on the *Beagle*, Darwin was shocked to witness the aboriginal peoples of Tierra del Fuego—a sight the likes of which most of his contemporary churchmen and naturalists had not seen. The experience humbled him, in the best sense of that word. He thought deeply and sensitively about his theory of human origins, ultimately concluding that it was "more humble" to acknowledge that both he and the "savage" peoples he'd seen were all descended from animals. Still, admitting to his belief in evolution felt to him like "confessing a murder" right up until the publication of *On the Origin of Species* when he was fifty years old. Yet he could not help but make his confession—he worked tirelessly day after day in the years leading up to that publication, striving to reveal the truth. As his journals amply indicate, if he had found reliable evidence in favor of a different conclusion, he most certainly would have accepted it.

In his two great works of science, *On the Origin of Species* and *The Descent of Man*, Darwin went out of his way to couch his ideas in terms as religious as he could possibly muster without completely diluting his scientific conclusions, both because of his own fears of Spinoza-style ostracization by his religious community and his sympathy for his devout wife, Emma, and their children. But afterward he turned to an autobiography intended for his family and for posterity and had the opportunity to write honestly about all he had discovered and pondered in half a century of exploration. In this work he talked clearly about being an agnostic, and about his devout commitment to a moral, noble life—"I feel no remorse from having committed any great sin . . . I believe that I have acted rightly in steadily following and devoting my life to science."[20] Through this commitment to science and conscience, truth and virtue, Darwin transformed society peacefully—carrying out a largely bloodless if not always painless revolution. In today's terms, he was a true Humanist hero.

The other original "new atheists" were more mixed in the degree to which their lives were exemplary, but each made a huge impact on Humanism. Marx, born in Germany in 1818, became an outspoken atheist in his early twenties, and though he too was spared a banishment like Spinoza's, his heresy did preclude a career in academia. So Marx concerned himself, for all his mistakes and despite whatever disagreements we may have with him ideologically, with an emerging phenomenon that urgently needed attention: the rapidly transforming global economy and the often desolate lives of its workers.

Influenced by the post-Christian German theologian Ludwig Feuerbach, Marx came to see religion as a projection of human hopes and needs onto a supernatural realm that did not exist; he insisted that we transcend all such projections, overcoming our inadequacy as religious believers as part of overcoming our inadequate working conditions by means of a Communist revolution. He possessed two of the important characteristics of Humanism: he was skeptical of revealed religion, and he was in search of a better life for all people. But Marx was also among the first great examples of the danger in assuming that, absent a perfect religious salvation, we can or should ever hope for a perfect secular salvation. There are no utopias. No utopian vision, godly or godless, must ever be allowed to justify violent repression or the coercion of conscience, no matter how noble the ultimate goals. Marx should be considered a precursor to Humanism and an influence on it, but Marxism is not and never has been the equivalent of Humanism.

Similarly, Nietzsche's powerful call that "God is dead" still sparks debate today. But his tragic life of mental illness and isolation is hardly the sort any Humanist would want to emulate. And in rejecting Christian morality, he made a number of statements that are difficult to accept. His work is dense, multilayered, and cannot be read in sound bites; I won't attempt to present any here. I mention him because of the surprising belief among some religious leaders and theologians today that when we talk of Humanism, we are speaking of Nietzsche's system of thought. No, he cannot be entirely absolved for the fact that some of those sound bites suggest, whether he meant them to or not, that the strong and fortunate have no moral obligation to the weak and powerless. Hitler and the Nazis took Nietzsche's work out of context and bastardized its meaning—but that is hardly high praise. In the end, his contribution to Humanist thought was mixed, and we won't debate it here other than to say that we *can* take from it his passionate insistence that any form of goodness must come without God, and any meaning our lives are to have must be a meaning we create.

Sigmund Freud was born in Moravia in 1856—three years before the publication of *On the Origin of Species*, making him the only one of our "Four Horsemen" to have been primarily a figure of the twentieth rather than the nineteenth century. And for all the outsized impact he had on that century as a brilliant—and certainly atheistic—intellectual revolutionary, in his case too we must hesitate before identifying his work or ideas too closely with contemporary Humanism. Freud did more than almost anyone to pioneer the scientific study of the human mind. But he worked so creatively, at such an early stage of that project, that he was bound to do a number of things we would consider grave errors if done today. His early flirtation with cocaine; his rigid insistence that psychoanalysis function like a kind of cult, not incidentally to promote his own voice and views; and his fixation on childhood-related sexual neurosis, with certain beliefs like the universal centrality of the Oedipus Complex to the human psyche as "shibboleths" separating insiders from outsiders—all these and more are hardly the stuff of which good contemporary psychology (or Humanism) is made.

Nonetheless, Freud's analyses of the origins of our belief in God and our need for religious authority taught the emerging secular and Humanistic world to deal seriously with religious beliefs and practices—to understand their origins and, most importantly, to respond in new ways to the human struggles and problems that our religious behaviors are attempts to work out. And Freud's vision of a mind divided into Ego, Id, and Superego contained within it the seed of a powerful idea Humanists must still reckon seriously with today: namely, that we are not simply the sum total of our conscious thoughts. We cannot possibly behave or think "rationally" all the time, because many if not most of our thoughts take place at the subconscious level, and much of our behavior and emotion is set automatically by our lizard brains before the brain's more recently evolved frontal cortex gets an opportunity to evaluate it. If we are to consciously choose the good life more often—with or without God—we must do the difficult work of exploring our unconscious motivations and learning to resist our most destructive instincts.

Neither Silence Nor Transgression

If Freud, Marx, and Nietzsche today seem like disappointingly incomplete symbols of Humanism as goodness without God, one often-ignored source to which we might turn for inspiration is the huge body of women's writing on Humanism and free thought to be found in Annie Laurie Gaylor's massive volume, *Women Without Superstition*. As Gaylor reminds us, the historic treatment of women by almost every major world religion is reason enough to seek out the good without God today. Contemporary liberal Christian and Jewish women have to do a lot of work to avoid or explain away Biblical passages like these:

Unto the woman he said, I will greatly multiply thy sorrow and thy conception; in sorrow thou shalt bring forth children; and thy desire shall be to thy husband, and he shall rule over thee. (Genesis 3:16)

Thou shalt not suffer a witch to live. (Exodus 22:18)

But I would have you know, that the head of every man is Christ; and the head of the woman is the man; and the head of Christ is God . . . For the man is not of the woman; but the woman of the man. Neither was the man created for the woman, but the woman for the man. (I Corinthians 11:3, 8–9)

Let the woman learn in silence in all subjection. But I suffer not a woman to teach, nor to usurp authority over the man, but to be in silence. For Adam was first formed, then Eve. And Adam was not deceived, but the woman being deceived was in the transgression. (I Timothy 2:11–14)

For those of us raised in late twentieth-century America, which has produced female secretaries of state, Speakers of the House, and presidential candidates, it is a shock to read such statements. Thankfully, women have been responding forcefully to the letter and the spirit of such material for hundreds of years now, with a positive focus on improving life for women and men in this world, here and now. And it can hardly be considered a coincidence that many of the most influential suffragists and later feminists in history have been Humanists. As just one example of dozens, Frances Wright, who according to Gaylor was the first woman in America to be the main speaker on a public occasion, made a well-known career as a free thought lecturer in the first half of the nineteenth century, and had the following as her credo: "I am neither Jew nor Gentile, Mahomedan nor Theist; I am but a member of the human family, and would accept of truth by whomsoever offered—that truth which we can all find, if we will but seek—in things, not in words; in nature, not in human imagination; in our own hearts, not in temples made with hands."[21]

But the singular feminist Humanist heroine, at least until recent decades when we have once again seen a flowering of many leaders, was Elizabeth Cady Stanton. Stanton was the first person to call for women's suffrage in the United States, and the author of the Nineteenth Amendment guaranteeing women the vote. She was a brilliant writer and a fierce and deeply patriotic advocate for equality. And she was the very model of Humanism in both her thorough, tough-minded insistence that this is the only world we have, and in her total devotion to making that world the best place it could be for women and men. As Gaylor highlights, Stanton was the most recognized female leader of her era, but has since been overshadowed by Susan B. Anthony, her less radical (but also nontheistic) contemporary. Though they shared many beliefs, Stanton spoke out louder than Anthony against religion as a force for repression of women's rights. She has paid the price in posterity, but we still have the opportunity to hold up her example in the way she held up that of Lucretia Mott, a liberal Quaker shunned by many as an infidel:

> I found in this new friend a woman emancipated from all faith in man-made creeds, from all fear of his denunciations. Nothing was too sacred for her to question, as to its rightfulness in principle and practice. "Truth for authority, not authority for truth," was not only the motto of her life, but it was the fixed mental habit in which she most rigidly held herself. It seemed to me like meeting a being from some larger planet, to find a woman who dared to question the opinions of Popes, Kings, Synods, Parliaments, with the same freedom that she would criticize an editorial in the *London Times*, recognizing no higher authority than the judgment of a pure-minded, educated woman. When I first heard from the lips of Lucretia Mott that I had the same right to think for myself that Luther, Calvin and John Knox had, and the same right to be guided by my own convictions, and would no doubt live a higher, happier life than if guided by theirs, I felt at once a new-born sense of dignity and freedom; it was like suddenly coming into the rays of the noon-day sun, after wandering with a rushlight in the caves of the earth . . .[22]

Can there be a better example of Humanist thought in action?

This takes us approximately to the late nineteenth and early twentieth century, when the word *Humanism* came into more widespread use in its meaning as goodness without God. There are many more stories to be told, many more great thinkers and leaders to cover. But again, my purpose here is less to look comprehensively at the Humanism of the past and more to respond to the desperate need for more Humanism in today's world, and in tomorrow's. We'll look at some of the more recent history of Humanist philosophy in the next two chapters as we discuss Humanist ethics; and in the second half of the book we'll explore some of the ways atheists, agnostics, and the nonreligious have recently come together to create positive, Humanistic alternatives to religion.

But for now, if you or someone you love is among the world's billion nonreligious people, the message to keep in mind is: You are part of a proud tradition, a world tradition. You are touched by the past and by your own ethical commitments, and you can touch the future on behalf of those who preceded you. You have ancestors from every corner of the earth who stood up for what they believed, and now you have a chance to take up their work and take it further. What are we if we forget their teachings? And what are we if we accomplish nothing new? If we build nothing new?

What is the next chapter in Humanist history?

CHAPTER THREE

Why Be Good Without a God? Purpose and *The Plague*

Why do you yourself show such devotion, considering you don't believe in God?

—Jean Tarrou to Rieux, *The Plague*

The Plague

In the small town of Oran, Algeria, dead bodies are multiplying exponentially. A strange virus has penetrated the town walls; it is causing people's flesh to boil, their insides to curdle with fever and vomit. There seems to be no hope for a cure. Terror has taken over.

One doctor, a darkly handsome man named Rieux, may be the last hope the people of Oran have—or he may simply have lost his mind. He is working tirelessly to treat the victims. He not only puts himself in contact with the deadly contagion, he does so methodically, with tremendous energy and unflinching dedication.

Admirable: but why? Rieux, openly an atheist, is confident he will receive no reward for selflessness after he dies. Furthermore, he reflects, even if he should succeed in curing the plague against all odds, all his patients and he himself will eventually perish. There will be no resurrection. All is temporary.

The doctor's beloved wife, meanwhile, is stranded in a sanatorium a hundred miles away. Oran is under strict quarantine. Perhaps his strange dedication to his patients stems from the faint hope that if he cures them all, he can be reunited with her? No—if this were all there were to it, he would not help his friend Rambert the way he does. Rambert, a journalist, had recently come to Oran for a visit when the plague erupted, and its quarantine has trapped him there. With suffering and death all around him, Rambert can think of little but escaping to reunite with his own young bride, in Paris. He is self-righteous in his longing to break free. He is not a citizen of the town, just an accidental victim of fate. He *deserves* to get out, and he is willing to break the law and put others at risk to achieve it.

Obviously Rambert misses the point. Unless the plague is some purposeful, vengeful work of God—and the thoughtful, skeptical journalist is hardly the kind to believe such fairy tales—the residents of Oran, screaming and dying all around him, are no less accidental victims than a tourist such as he. *We are all accidental victims.*

Still, Rieux does not resent Rambert's choice of selfish love over selfless service. The doctor does not attempt to persuade his friend to stay, despite the desperate need for more help in the "sanitary corps" that Rieux is organizing. Rieux encourages Rambert to follow his own heart. If the doctor were working for others only out of raw, calculated self-interest, surely he would calculate the need for Rambert's participation and be less understanding. Instead, Rieux simply persists in an endeavor that cannot but inspire us, despite describing his own struggle as "a never ending defeat."

The philosopher Albert Camus's novel *The Plague* dramatizes one of the most foundational and challenging questions for Humanism. We nonreligious people may easily be able to answer our most obnoxious critics—those who cannot muster even the intellectual self-restraint to refrain from suggesting that godless people cannot be good and thus must be some lower form of person. But let us recognize that most of our religious critics, or those who would call themselves skeptical of our particular brand of skepticism, are more sophisticated, not to mention more good-natured. For their benefit, and for our own, we must explain *why* we should be good, without God.

This is the question we can hear the character Jean Tarrou, a little bit admiring and a little bit ashamed, asking his friend Rieux: "Why do you yourself show such devotion, considering you don't believe in God?" It's hard to overstate the importance of this question. Because to answer it requires much more than simply a *retort* against the famous line written decades earlier by Fyodor Dostoyevsky for his character Ivan Karamazov: "If God is dead, all is permitted." Camus demands—and Humanism ought to provide—more than simply a *rebuttal* of the notion that we should eat, drink, and be merry, for tomorrow we may die. We must get beyond defensiveness. To paraphrase contemporary civil rights and green movement leader Van Jones, Martin Luther King Jr. did not get famous for a speech entitled "I have . . . a list of snappy rejoinders!" So this chapter will be an examination of the different ways Humanists and the nonreligious engage with another old, familiar question: "What is the meaning of life?"

Of course, when we talk about "the meaning of life," many intelligent people scoff at how self-centered the phrase may sound. Focusing on the meaning of *my* life seems to be all about, well, *me*. What if the meaning of my life has nothing to do with the meaning of *your* life? What if I discover that the meaning of my life lies in exploiting you? Indeed, the meaning of life is only really worth thinking about if we start with the premise that a meaningful life is a good and ethical life. Humanism is an acknowledgment that a meaningful life is by definition a moral life, and a moral life is by definition a meaningful life.

But we will also discover that morality is not about sinners and saints, heaven and hell, damnation and punishment. It's about alleviating unnecessary suffering (some suffering is necessary) and promoting human flourishing, or dignity. But before I try to convince you that the meaning of life is dignity, as opposed to, say, the number "42" (the answer given by a giant supercomputer built solely to answer this question in *The Hitchhiker's Guide to the Galaxy*), let's explore further why we would even care about the question of life's meaning, and what some other possible answers to that question might be.

Purpose Driven

Most atheist leaders and spokespeople are good at telling you why we should teach evolution and not creationism in schools, or why Jesus couldn't possibly have been born of a virgin and resurrected from the dead.

But the litmus test for whether such leaders are worth much, to me, comes in whether or not they've got anything helpful to say about this feeling Camus describes—the way that, even when we're not coping with the bubonic plague, life can sometimes seem meaningless, like a "never ending defeat." I was flabbergasted to read Richard Dawkins's blog about his response, during his 2006 book tour for *The God Delusion,* to a young man who approached him and asked, "Dr. Dawkins, I'm thinking of committing suicide—what do you have to say?" Dawkins could at first think of nothing better than to suggest that the kid put his question to the readers of the discussion forum on the Richard Dawkins Foundation Web site. Then Dawkins got a brilliant idea—if the young man had been at Harvard, he could have gone to the Humanist chaplain, or if he'd been at Oxford, he could have visited the Anglican chaplains, some of whom, in Dawkins's words, are "very nice people."

Is that the best we can do? Rage, rage against the dying of the Enlightenment, then shoo our troubled youth right back to religion because we're too distracted or cerebral or both to spend a few minutes of our deep thoughts on how to be more loving, more helpful? I admire Richard Dawkins and am thankful for the majority of the work he does. But that essay of his stands as part of the public record, and it is troubling, because Dawkins has been cast as something of a world atheist spokesman in recent years, and he can be emotionally tone-deaf on this crucial issue.

In past generations, leading voices for Humanism and the nonreligious perspective were more articulate in their response to the human search for purpose and meaning: Freud, Camus, Abraham Maslow, Erich Fromm, Sherwin Wine, and others come to mind. But every generation must discover these issues anew, especially the current one, which most likely includes the highest percentage of nonreligious people of any generation in recorded history. Still, we've yet to see many concrete gains from that, because we haven't discovered the common purpose, the mission that will unite us as a movement.

Ironically, in recent years the author who has best understood the human drive for meaning and purpose is the Reverend Rick Warren, pastor and author of *The Purpose Driven Life.* Warren's book has been such a huge success because he has recognized our need for purpose beyond self-centeredness and because he has seen that many nominally religious and nonreligious people are struggling with a sense of purposelessness today, even as they gain in both scientific knowledge and material prosperity. "The man without a purpose is like a ship without a rudder—a waif, a nothing, a no man,"[1] as Warren quotes Thomas Carlisle, clearly touching a chord with tens of millions of readers. And indeed, as he points out, we need to *choose* our own purpose, our own drive, because otherwise we will spend our lives *driven* by any number of subconscious drives programmed by our selfish genes: guilt, resentment, anger, greed, fear, the need for approval, etc. Think of this list and ask yourself, if not about yourself then about the people around you—how many of them are slaves to one or another of these? And are you, more often than you would like to be? Warren argues compellingly that "It's not about you," insisting that life and society are at their best when we overcome selfishness and solipsism, learning to live for a purpose higher than ourselves.

Unfortunately, Warren's idea of what life's purpose ought to be is unacceptable, rendering his *Purpose Driven Life* a startling combination of the sublime and the abominable. He writes with crystal-clear confidence that life's higher purpose can only be Christianity, as interpreted by people like him, and that those who disagree will spend eternity in hell, "apart from God forever."

Again, Warren has repeatedly called atheists arrogant, proudly admitting on national television that he would vote for anyone *but* an atheist for president. But it is the height of arrogance to be so openly prejudiced against those who agree with him about the need for purpose but prefer Islam, Judaism, or simply the Humanist faith in our ability and responsibility to build a healthier world for the sake of our loved ones and all humanity.

A Word for God

But first, what about God? What is so *wrong* with the idea of God as a motivation, as the way for us to understand our purpose in life?

Well, if it really, truly does motivate you to be good, then nothing. I have no quarrel with you. Again, this book is neither an attempt to convert you nor an attempt to debunk your purpose.

Of course, part of any friendship, and any respect worth offering, is the mutual freedom to express ourselves and state plainly where we disagree. True friends don't go around insulting one another: "Bob, you're ugly, whiny, cross-eyed, and I can't stand the sight of you!" But a friendship where we can't share why we passionately disagree is not a friendship worth having. I for one was not raised at a dinner table where we put off all discussion of politics and religion for the sake of genteel politeness, and I don't expect the readers of this book, whether religious or not, to need that kind of intellectual coddling either.

That said, a glance at the quote above from Warren reveals that, for all his insight, success, and credibility, he can at times be no less infuriating than a sidewalk preacher on a university quadrangle, harassing and haranguing passing hipsters with threats of hellfire and damnation.

When I used to see such preachers as a college student, I often couldn't resist the temptation to tell them: if the meaning of life is to accept entry into a heaven where most of my ancestors and most of the people in the history of the world could never go, simply because they hadn't heard of or didn't accept his God—well, that wasn't a heaven for me, with all due respect. If the only way to get to heaven is by worshipping a God who had the power to prevent the Holocaust but chose not to, no thank you. But if God was *not* capable of preventing a tsunami, or a genocide, or even small disasters like the death of a spouse or a child or the loss of a limb, because only people can prevent those things, then why would we need God? If humans are on their own in a universe that is either mismanaged or not managed at all, isn't it better, at least for some of us, to have the integrity to admit that fact and begin attempting to manage our own lives and societies as best we can? And really, even if we were to accept Warren's point that only God's purposes are worthwhile, we get nowhere. Because we still need to rely on human beings to tell us *what God's purposes are*. This is what has caused so much of the religious war and sectarian hatred—people who agree wholeheartedly with one another that we must follow God's purpose but slaughtering each other over sometimes tiny differences of opinion about what God said his purpose happened to be.

Actually I am grateful for Warren's extreme language because it provides an opportunity to say to moderate self-identifying Christian readers, or to readers who may say they aren't religious but do believe in God or just aren't sure that they're Humanists: it's not an accident that Warren and people like him feel so justified in portraying themselves as the "authentic" voices of religion. This definition of God, as a real being that can really damn and punish you for all time, is the only kind of God worth dying for belief in. If you believe that's what God is really like, by all means, worship God. But if you believe God is nature, or love, or the universe, or "ultimate concern," or "that which is larger than us," do you really think Rick Warren would consider there to be any difference in your belief system and Humanism? He wouldn't; and it raises the question—why should you?

There are those who are truly motivated to be good by terror of God's supernatural punishment and hope for his miraculous rewards. But maybe if you are among the many millions who *don't* literally believe in heaven or hell, then you too are a Humanist. Because if the purpose of your life isn't to avoid hell or get to heaven, then as much as you might like to say it is to "love God" or anything like that, you will have to *choose* a purpose based mainly on how you as a human being should relate to other human beings in this world, for the sake of this world. Just as Humanists do.

Humility

Along with belief in God, two central values of human history have been humility and submission. Life was so unrelentingly difficult for so much of humanity for so long that it didn't make any sense to strive to improve your conditions. It was like beating your head against a brick wall—pointless. *Submission* was almost the only way for a sane person to respond to a situation in which there was no way for common people to figure out why it would rain for weeks and destroy their crops, then go dry for months and starve them some more. We had no answers for why so many of our babies died in childbirth. We were oppressed by armies impossibly more powerful than we were, and there was nothing like representative democracy to invoke in response. I could go on. What you got in a life like this was suffering, a little pleasure here and there, close companionship with others around you suffering similarly, more suffering, a little more pleasure, and then probably some more suffering before you died. Much of our modern culture would not have made sense to people living this kind of life.

For example, think of Nike's famous slogan "Just Do It." The serf or common man in ancient times would have been bewildered by this message. "Just do *what*?" There was nothing they could do to escape the harsh reality of their fate. And so a key value of many ancient religions is humility, which is really the ability to tolerate suffering by accepting it as both deserved and good for you.

Humanists acknowledge that there is still a great deal we can't change about the conditions we live in. But we look all around us and see how much we can change and how much we have changed. We might occasionally have to resign ourselves to our fate. But more often, we value the ability to push forward with ambition, vision, and a sense of humor when things get tough, instead of humility, resignation, and submission.

Life's a Bitch, and Then You Lie (to Yourself)

For those who have dispensed with the idea that there is a God that created or creates life's meaning, a first and ever-available response may be nihilism—the idea that life simply has no meaning, and can have none. In fact, we Humanists, atheists, and the nonreligious are often forced to defend ourselves against charges that our worldview, lacking God, is nihilistic. Is this a fair accusation? No. But not because no atheists are nihilists. Rather than just echoing the most jingoistic, patriotic kinds of religion when we talk about ourselves as nonreligious people, let's have the integrity to admit that of course it is possible to be a nonreligious person who consciously or unconsciously adopts an attitude reminiscent of the worst stereotypes of godlessness. But it is equally possible to give lip service to being devoutly religious—maybe wearing a fancy cross around one's neck, or ostentatiously attending the best temple and sitting in the best seats. A recent trend among Muslim men is competing with one another over who has a bigger *zebiba*, or bruise on the forehead from prostrating himself in prayer.[2] There are plenty of people who love these kinds of religious displays but actually couldn't care a fig about anything but themselves.

Nihilism is a real problem in this world, and it afflicts the religious and the nonreligious alike.

But many people claim that Humanism or atheism *is* nihilism, and vice versa. This assertion is either unconscionable, incredibly ignorant, or both. So what is the concept of nihilism, really? There are several types.[3]

RUSSIAN NIHILISM:

The word *nihilism* began to appear in late nineteenth-century Russia but was popularized in novels by Turgenev and Dostoyevsky that suggested that, in a world without God, there is nothing to embrace but reckless hedonism. In the novel *Fathers and Sons*, Turgenev's character Bazarov declares that he is a nihilist, which to him means to "act by virtue of what we recognize as beneficial."[4] Bazarov then defines "beneficial" as to deny and negate . . . everything. No values. No worthwhile morals. Nothing is true or good.

Dostoyevsky imagined that adopting this view that life can have no purpose would seem to end logically in murder, suicide, or other extreme forms of hedonism—after all, if there are no values worth adhering to, then there's no good reason why we should be kind to people if kicking them into the mud should happen to provide us with more obvious entertainment. There's no reason to spend years studying medicine, a pursuit requiring agonizing self-discipline, if going around killing people and stealing from their corpses should happen to provide more of an adrenaline rush. Clearly, concerns about this kind of nihilism are still with us, because a contemporary symbol of "Russian nihilism" is Batman's nemesis, the Joker. The Joker is a sociopath who believes that life is little more than a cruel joke and who is prepared to act boldly on his belief by shamelessly exploiting others for pleasure. Which is precisely why we need Batman. Some people may think this kind of anything-goes nihilism is Humanism, but they are wrong.

SCHOPENHAUER'S NIHILISM:

This is a more depressed, inward version, often flippantly but not inaccurately summed up as "*Life's a bitch, and then you die.*" Schopenhauer, a nineteeth-century philosopher, didn't see any positive meaning in human existence and in fact believed the universe might actually be hostile to humanity, given how much suffering he saw it heaping upon people. Yet he did not seem to derive any great pleasure from inflicting misery on others.

The Schopenhauerian nihilist would hardly be a likely psycho killer like the Joker; picture *Winnie the Pooh*'s Eeyore, always moping and miserable even on a day when the sun is shining, the birds are chirping, and others are trying to smile at him. This is not necessarily such a menacing person, but if he could appreciate life more and enjoy it more, if only to be able to laugh at himself a bit, he'd clearly be better off—and so would you be if you happened to be the person in the next cubicle. Schopenhauer's nihilism is not Humanism either.

THE NOBLE LIE:

This third type of nihilism is perhaps the most popular. The phrase was popularized by contemporary writer Loyal Rue, who argued in *By the Grace of Guile: The Role of Deception in Natural History and Human Affairs* that there are no true values worth living by, but that if we lie to ourselves and say there are, in fact creating entire elaborate moral and social systems based on these lies, things will go much better for us. This might sound idiotic at first, or at least far-fetched, but it is a significant theme in contemporary philosophy, so much so that philosophers have two different names for it: "fictionalism" and "error theory." In fact, error theory, or the Noble Lie, might be the most popular form of nihilism, because it allows people who suspect that there is really nothing worth living for to forget their suspicions for a while and live as though there were meaning in life. But you have to wonder if having in the back of your mind that you are living a lie doesn't have at least some pernicious effects on a person.

Astonishingly, this concept recently seemed to have leaped off the pages of musty philosophical tomes and actually taken over our national and global political lives for almost an entire sordid decade. A famous example of an influential error theorist or "noble" liar would be the philosopher Leo Strauss, the former University of Chicago professor who inspired most of the important contemporary politicians known as "neocons," such as Paul Wolfowitz, Donald Rumsfeld, and Dick Cheney. Strauss did not believe religion to be literally true, but felt the American people needed heroic myths to follow and be ennobled by. And so, Strauss suggested, we must give them those myths—including by playing up the differences between our civilization and other civilizations.[5] And if a loosely justified preemptive war on foreign soil should follow? Well, then, just consider that "collateral damage" in the struggle to provide the myth that will help stave off the truth of nihilism.

What all three of these forms of nihilism have in common is that they all deny any religious meaning or purpose to life, but they are also not Humanistic. Each denies the possibility of any *genuine* secular, Humanist, or nonreligious purpose or meaning.

While it is possible to give up religion in favor of one or another form of nihilism, the vast majority of nonreligious people are neither nihilists nor believers in any supernatural or absolute values. As Humanists, we are what some philosophers call "subjective realists"—we know there really are such properties and values as good, fear, pain, and meaning, but they are dependent on humans for their existence. In her book *The Second Sex*, Simone de Beauvoir describes femininity in these terms. Femininity would be nothing without people's beliefs, she argues, but this doesn't make it false—there are women. From this point of view, focusing a lot of attention on the fact that there is no objective framework of meaning imposed by the universe is little more than an attempt to escape responsibility. There are many meaningful and difficult choices we must make every day of our lives, even if the universe doesn't spell them out for us in giant stone tablets floating in the ether.

Disbelief in God does not imply nihilism. So what are the subjective purposes nonreligious people might be driven by? And which of them are really worthwhile?

The Strivers

I'll never forget the time one of my fellow Harvard chaplains, a southern Presbyterian and all-around great guy named Brad Barnes, took me aside at a breakfast meeting and said that he'd always wanted to ask me how it felt to be a chaplain for the largest denomination in the world. At first I had no idea what he was talking about. Surely he was aware that there were individual Presbyterian churches with more members than the entire American Humanist Association. But after a bit of discussion, I figured out what he meant.

Most people in the world say they believe in God. But whether they do or not, faith in God is *not* what drives the vast majority of them. If you added up all the nominal Christians, Jews, Muslims, Hindus, Buddhists, etc.—those who are religious in name only, who do not take the tenets of their religion very seriously—you really might get the largest denomination in the world. These are people who might not be self-consciously Humanist, who might consider themselves religious, but have been secularized. As Sherwin Wine liked to point out, they'd much rather play golf on Sunday or go to a movie on a Friday night than go to church or synagogue. From the point of view of a minister like Brad who really cares about engaging his flock with the life and theology and community of Jesus Christ, these people might occasionally set foot in a chapel, but they are far, far outside the tent of true religion.

What I had to explain to Brad was that being a nominal Christian or whatever does *not* necessarily make you a Humanist. If obeying and loving the Christian God is not what's driving you, it doesn't mean that Humanism *is*. As I've been arguing, you may in fact be closer to being a Humanist than anything else, but you still have to take responsibility and *decide* what you are. I can't do that for you. No one else can.

I suspect that what drives most people in this world, what really might be the world's largest religion, is one form or another of what we can call "striving." Do you know any strivers? These would be people for whom the meaning of life, whether they admit it or not, is *get*, *get*, *get* as much as you can. Get really rich, get really powerful, get the nicest cars, the nicest babes, the nicest jobs, whatever. (Of course, god help you when this is the only way you know how to live, and suddenly the market crashes into a Depression, right?)

Actually, striving is a variation on Schopenhauer's "life's a bitch" theme, but instead, life's a race. If people who think life's a bitch and, understanding themselves as nothing but victims of unfair circumstance, tend to sit around and mope, then those who believe life's a race are the ones always working, working to win the race. Sometimes winning a race can be exhilarating and inspiring. But if our entire lives are driven by little more than the desire to win—to acquire things or people or status? As Lily Tomlin famously said, "The problem with the rat race is that even if you win, you're still a rat."

It's not hard to understand how a truly pious person, genuinely making an effort to live a life of religious modesty and compassion, could become frustrated and angry about the constant striving and selfishness we see in certain areas of our popular culture. I love high-quality pop music, for example—I see great rock, folk, and rap songs as today's secular alternative to liturgical prayer—but the majority of pop is little more than a catchy hook and a drumbeat browbeating us to think only of ourselves and what is in life for us—maybe sex, maybe a little love, maybe "Money, Cash, Hoes," as the Jay-Z song goes.

Forgive me if I mention a hip-hop song to make a serious point about religion, but I am a member of the first generation of Americans to grow up immersed in a pop culture that felt more like religion than actual religion. And I could never put my finger on exactly why this bothered me, until the morning of September 12, 2001.

I'd spent the first half of September eleventh trying to call my mother, who lived three blocks from the World Trade Center, to find out if she was alive. I was at the University of Michigan Hillel (the campus center for Jewish students), where I had gone for a meeting about the Humanistic Jewish student group I led, and I was surrounded by fellow students also worried about their loved ones. The second half of the day was spent working feverishly alongside dozens of fellow student leaders, university officials, and the university psychological and counseling services office to put together what became a beautiful candlelight vigil attended by twenty thousand people. We all had a sense of common purpose, and there were many tasks to do together over the course of only a few hours. I remember clearly that there was no music all day—only the sound of our own voices, often at a whisper, or of the latest radio and TV news updates, or of a silence thick with purpose, until we all sang together in unison at the vigil.

The next morning I awoke, surprisingly not so much terrified by the horrific events of the morning before as energized and inspired by the selfless community-building work we'd done the evening before. But then I walked into a bagel shop to get breakfast, and the radio had been turned back on: the first music I'd heard in twenty-four hours was the rapper Nelly's "Ride with Me," a song whose protagonist proudly invites his listeners to join him in cruising around in a four-wheel-drive truck with solid gold rims, and then smoke with him in the backseat of a Mercedes. The song's painfully catchy chorus chants again and again, "Oh why do I live this way? Hey—it must be the *money*!" It must be the money?

I don't know if I'll ever forget the feeling of disappointment and outrage—*this* was our response to the assault of fundamentalist Islam on our values? The values Nelly's song and so many like it speak to—striving for the most money, the most women, the most pleasure or status or power or drugs—were *these* our values? Don't get me wrong: if I wasn't familiar with this kind of music, if I didn't enjoy a lot of it, I would never even have been able to make out the words. The song, like so many others like it, is extremely catchy and a lot of fun to listen to if you pay no attention to what it's saying. And of course I value living in a secular country where religious or totalitarian authorities can neither censor nor forbid such messages officially. Furthermore, much of rap music is not at all shallow. But hearing this kind of song in this kind of setting was a powerful reminder that a secular culture is not the same as a Humanist culture, and that sometimes the former falls far short of the latter.

Still, part of the lasting appeal of hip-hop music is the theme of the underdog who makes good. Even the most materialistic hip-hop often has an element of celebration that those who might have been the victims of circumstance or oppression have overcome, made it big. Which raises the issue: obviously there are some rats who seem to win the rat race with such glory, such distinction as to give lie to Camus's idea that life feels like a never ending defeat. Some people seem to strive ambitiously toward some goal and achieve it beyond their wildest dreams. Some people seem to be happier, to live lives more charmed than the rest of us. Julius Caesar was one of the earlier archetypes for this kind of person.

Of course, we don't have a very detailed or objectively historical picture of Caesar's life, but then again, isn't it a common feature of those we idolize as rat race champions that their lives tend to be shrouded in a little mystery, beyond reports of their great victories? Once we get to know more about people, from Princess Diana to Britney Spears to Bill Clinton, even the brightest celebrities tend to disappoint. Caesar, Montaigne relates in his essay *The Story of Spurina*, was a man who possessed many tremendous virtues. He was unparalleled in energy, eloquence, and, when it did him no harm, magnanimity. But the emperor's highest purpose in life was his own "furious passion of ambition," and he ultimately squandered the might and good name of his own republic, to whose defense he was sworn, to sate his many lusts. Ultimately allowing himself to be worshipped as a god in his own presence, Caesar's example should give us pause any time we set out to choose the meaning of our own lives. What will stop us from becoming, in our own fashion, the equivalent of him? Ambition is a healthy part of life if set in the service of worthy goals. But when it becomes the entire point of life, there is no way to satisfy it, because there will always be another battle, another potential conquest. When is enough *enough*?

The more common striver, though, is the person who works and works and works toward a goal he can never attain, or who attains it and then realizes: I can strive all I want but it's to no end; all is vanity. These are stories I hear all the time: I worked all my life to get a great job. I got the job and I was bored, it didn't fullfill me . . . I worked for years to get the best education. I studied for years and years, and when I graduated I wasn't sure if I was any wiser than when I began . . . All my life I worked to find the most attractive partner, then I found someone who seemed perfect, I won her over, and now that we're married I'm not attracted to her anymore.

This is why we could talk of nothing else when the story broke about the downfall of Eliot Spitzer, former governor of New York. Spitzer had the perfect job, the perfect life. He was a young, popular governor, potential future presidential or vice presidential candidate, with a reputation for busting bad guys and cleaning up scum. He had a beautiful, accomplished wife and beautiful children. And he paid money—thousands of dollars an hour, to be specific—to a young prostitute to help him throw it all in the trash.

Spitzer became an astonishingly perfect example of a truth we all knew but couldn't allow ourselves to acknowledge: that no amount of successful striving can calm the emptiness we are prone to feel. When we are driven by nothing but victory, no victory can quench our thirst for yet another. One of the biggest reasons why millions of nominally religious and even nonreligious people turn to Evangelical or Pentacostal or Wahabi or other strict and conservative versions of religion is that they rightly recognize that if you replace a religion like Christianity or Islam with little more than the worship of winning, you've gained nothing. No matter how much you win. And since most people are not aware of a secular and Humanist alternative to the life of striving, they often choose a return to a conservative religion that enforces its own goals and agenda, because no matter how unreasonable or sometimes even murderous those goals and agendas might be, at least they aren't solipsistic or isolating, and so we're not as prone to feeling hollow and alone inside when working toward them.

Can Striving Be Eliminated? *Should* It Be?

Buddhism, or Eastern mysticism, or simply "spirituality," whatever the last might mean, is often offered as a somewhat nonreligious alternative to the life of striving. It's easy to understand how people can get fed up with the constant struggle to achieve, and look to eliminate their desire for achievement as a way out of the struggle.

As a teenager, I used to think a lot about what you could call a generalized form of non-Western mysticism. I grew up in New York City with two nominally Jewish parents who were surprised, to say the least, when I told them that I wanted a bar mitzvah. I'd been going to some of my older cousins' ceremonies and it just seemed like a fun thing to do. Dutifully, they took me down to our local Reform synagogue, the Free Synagogue of Flushing. FSF, to my ten-year-old eyes, seemed like an imposing old structure with lots of marble and gold, a grand and towering dome, and elaborate stained glass everywhere. But the place had a palpable feel of mustiness from decay, as Flushing had for decades been morphing from an area heavy with Jewish immigrants to a kind of new Asiantown teeming with much more recent immigrants to the United States. The Free Synagogue and its Hebrew School were clearly experiencing this downturn, and the place felt anything but free to me.

The teachers seemed apathetic, the prayer doubly so—"Blessed art thou, oh Lord our God, ruler of the Universe, who sanctified us with his commandments . . ."—people were mouthing the words but expressed none of the awe or reverence that seemed as if it should accompany such words. As Sherwin Wine used to say, the mistake of the Reform movement was to translate the prayers from Hebrew to English. Once modern people could actually understand what they were supposed to be saying—mistake! Of course, I later learned that not all Reform synagogues are so old-fashioned, and some can be quite dynamic and rich with creative programming. I even came to see some beauty in the haunting chants of the Free Synagogue's cantor, Steve Pearlstein, though by that point I was convinced enough of my Humanism that no amount of socializing or music could have persuaded me to believe in a liberal religion. But that's another story.

In any case, I tried to get out of the bar mitzvah, but my parents forced me to go through with it, not for religious reasons, but to honor the commitment I'd made. A good lesson, but it meant that as soon as I was done, I found myself looking to get as far from Judaism as I could, and so I started picking up some of my father's many books on Eastern religion and mysticism. Before long I couldn't put these books down. In high school I read the *Tao Te Ching*, the *Baghavad Gita*, as well as modern texts like *The Way of Zen*, *Cutting Through Spiritual Materialism*, and Carl Jung's work on alchemy. I wasn't entirely sure how to explain this interest, but I felt I had tapped into something true, something most of the people around me just didn't see. In college I majored in Chinese and religion and worked harder than I'd ever worked to learn enough Chinese to be able to go to Taiwan and study Zen in its original form. Zen is actually just the Japanese translation of the Chinese word *Ch'an*, which comes from the Sanskrit *Dhya-na*, meaning meditation. Ch'an is known as one of the most rigorous, most esoteric forms of Asian religion, and these days it can only be found untouched in Taiwan because the Communist government in mainland China has rooted out most of it over the past century.

All this I learned from a white, culturally Jewish professor of Buddhist studies at Michigan named Robert Sharf who, though he seemed neither very interested in "spirituality" nor in meditating, was known to be an ordained Buddhist priest. I thought often that I too might spend my life as a Buddhist or Taoist priest—whatever that meant.

Then I got to China and realized that most Buddhists and Taoists are no more serious about their religion than the Reform Jews I'd known in New York were about their Judaism. Questions to Buddhists about their meditation practice were just as often met with a cynical shrug as with a serious answer. Ch'an meditators and Taoist holy men would place their little shrines next to a jug of wine or a TV/VCR and a couple of pornographic videos. I started reading more, and it dawned on me that not only do most Buddhists have heavens, hells, gods, ghosts, magic spells, and rain dances, but even the most "exotic" forms of Buddhism like Ch'an and Zen, or East Asian meditation Buddhism, can have very religious structures, a heavy focus on authority, fights over sects and lineages, and histories encrusted with legend and myth. And Zen meditation, because of the rigorous discipline of mind and body it promotes, was an extremely effective tool for training kamikaze pilots (perhaps the original suicide bombers) in World War II.

It's not that any of this made the Chinese Buddhists I encountered *bad people*. I loved meeting almost all of them and could not have been more grateful for the way many opened their lives and their small homes to me as I traveled across Taiwan and the mainland, where I did meet many who identified as Buddhist and Taoist, though rarely with much passion. But what struck me, wherever I went, was that although these folks were nominally Buddhist, they seemed to have absolutely no connection with the basic tenets of Buddhism—they had no fewer desires than anyone else, no more "equanimity," and they exuded neither more nor less "inner calm" than rural people you might meet anywhere else. People are people, I thought, and still think. Eventually I realized I'd gone to China to look for a way to get beyond all the fruitless, nonstop wanting that I saw all around me in New York and Michigan—people wanting the right job, the right education, the right sex, the right clothes. But people want just as much in China—if they didn't, they would never have had the motivation to win more Gold Medals than any other country at the 2008 Olympics—which is why, despite all attempts to impose socialism, free markets and an enlightened capitalism will always be a better fit for the Middle Kingdom.

Maybe I had been susceptible to the way all Eastern and South Asian philosophical ideas tend to be packaged and marketed in America as the spiritual equivalent of an exotic Miss Universe: she's brown, she's mysterious, she's soulful, she's out of your league—you've got to have her. The idea that the purpose of life is to achieve oneness with the universe, or to transcend it, is essentially the heart of New Age philosophy, which, we are taught, equals Buddhism, Taoism, and certain forms of Hinduism, though that's not really the case. It's more that New Age thinking is a recent, primarily Western invention that sometimes comes packaged as a form of Asian religion. This is to get past the obstacle that people prefer philosophical ideas that seem ancient and time-tested as opposed to new and risky, and prefer foreign and exotic ones to the local, familiar, and dull (been there, done that). Even the ancient traditions that actually do emphasize transcendence over all things, like authentic Ch'an Buddhism, tend to advocate achieving it exclusively by means of methods like days-long silent meditation at pain of flogging if one scratches one's nose, walking on broken glass, or other techniques that wouldn't go over as well with the neighbors, and so are generally glossed over for Western audiences.

Ultimately, New Age spirituality has its strengths and its weaknesses. It generally attracts well-meaning individuals and only rarely leads them to do anything actively harmful to other people. And New Age practitioners are sometimes taught to renounce emotional and material possessiveness.

This can perhaps go too far, however; as in the case of the late Tibetan Buddhist leader Chogyam Trungpa, author of *Cutting Through Spiritual Materialism*, about whom laudatory lectures are still given at Harvard by professors like the liberal Christian theologian Harvey Cox. Trungpa was known to ask his richer students, in the name of shedding attachment, to buy him a Rolex watch or a luxury car.

All this simply made it easier for me to decide that it's impossible to eliminate all desire, and even if it weren't, some desire and striving is healthy. The question is: What are we striving for?

Just because we have one almost completely undocumented story of a mythical prince in India having eliminated all desire twenty-five hundred years ago doesn't mean it can actually be done. In fact, it's amazing that if you take Buddha as the ideal model for Buddhism and Jesus as the ideal model for Christianity—they both seem to be based on real men, and both seem to have had some fascinating insights and opinions about life—there's no more evidence for either the existence of the Buddha or that he actually performed all the feats attributed to him than there is for the story of the Gospels. Yet many skeptical people, who would never give any credit to the Christian Gospels being literally true, like to assume that it is actually possible to be a living, breathing human being and desire nothing.

Besides, even if it were possible to want nothing, why would you want that? In fact, even most people who find Buddhism fascinating would probably admit, if forced to consider it, that there is something worthwhile about desire and passion and about *caring.* Just as people considering suicide often fantasize about how wonderful it will be at their funeral, hearing everyone miss them and finally appreciate them—not stopping to realize that they won't be around for any of it—much of our Western thinking about Buddhism misses the point that without desire and passion and caring, we wouldn't have love, we wouldn't have sex (Buddhist monks are supposed to be *celibate*, after all), we couldn't even have friendship as we know it. Friendship is a more relaxed form of desire, but nonetheless close friendships are so powerful because they too are fueled by our desire for companionship and appreciation.

Of course, it is certainly true that there are elements of Buddhism and New Age religion that are secular, realistic, and do real good for their practitioners. Just because the idea of eliminating all desires is unrealistic and undesirable doesn't mean there aren't some desires we'd be better off eliminating. Buddhist meditation techniques are among the tools available to us to help eliminate those unproductive desires and cultivate healthier attachments.

For Humanists, it is good to desire, and it is good to care. The questions are: what do you desire, and what do you care for? Humanism's message is no more or less than: be passionate about things that are *worth* being passionate about.

So what are these things that are most worth being passionate about? What are the purposes we can choose that are the best alternatives to either Christian or Buddhist or any other purposes, including the lie that we can live without any purpose at all?

What Are We Striving For?

All You Need Is Love

There are biological reasons why the biggest motivator for the largest number of human actions and decisions is our desire for romantic, sexual, and other kinds of love. We are profoundly social creatures—in order to fit our big brains into our big heads and still slip them through a narrow birth canal, we evolved to be born so helpless that for the first few years of our lives we have little chance of survival without almost constant love and nurturing from those around us. And because so much caring for children is needed, evolution programmed parents to find long-term love bonds with one another, to help keep them together for the long and arduous task of parenting. All of this took shape over the course of millions of years. We did not choose our need for social connections, or for sexual and romantic love, and we did not choose to evolve so that every instance of love would be so imperfect, every partnership match so inexact. But like it or not, we are responsible for what we do about it all in this life.

Much of religion has been about trying to find a solution to this constant, ever-imperfect hunger for companionship. Not coincidentally, every major world religion has strict laws regulating marriage, sexual purity, and fidelity, when men and women can and can't be around one another or touch one another. And much of our time and energy even today, in every society, is devoted to obeying these rules, rebelling against them, or trying to come up with new rules to replace them. Each society evolves its own way of working these challenges out, which results in conflict when different traditions come into contact. To give an ancient example about which you might not be aware: though Buddhism is now closely associated with China and the Confucian cultural landscape, when it first came to China from India it was strongly resisted because it challenged the notions of procreation and family bonds, encouraging men—and women—to sever ties and adopt a renunciate lifestyle that went against the Confucian traditions regulating marriage and child rearing. Even today, most Western fans of Buddhism don't realize that the Dalai Lama, as an orthodox Tibetan Buddhist, is deeply opposed to gay marriage.

Jonathan Haidt describes a "happiness formula" in which our happiness (H) equals our biological set point (S)—meaning the extent to which our brains are wired to allow for feelings of happiness, uplift, and joy—plus the conditions of our lives (C) and our voluntary activities (V), or $H = S + C + V$. Haidt argues that love (along with work) is the single biggest element of C.[6] To translate, for those who weren't expecting mathematical equations in this book: Love is not just some voluntary, extracurricular activity that we can pick up and put down when we please. And it's not some set or fixed biological reality totally predetermined by our genes to make us miserable or blissful, or both at the same time. The degree to which we have love is a fundamental condition of our lives, like the degree to which we have housing, clothing, money, education, or access to crude oil fields.

Just like any of the other fundamental conditions of life, if we don't have love, we may be able to get it with hard work—and if we do have it, we shouldn't get too haughty about it, because we can lose it at any moment. Granted, we may not be able to find love without some of the other conditions at least partially fulfilled—as the novelist Daniel Handler wrote in his irreverent novel *Adverbs*: "Who can fucking dare to tell me that love is intangible when it's so obvious that it's not? The people who say intangible have places of their own."[7] But assuming we have our basic material needs even somewhat satisfied, no single thing is going to take up as much of our time and energy as finding, holding on to, or wondering what happened to love.

But to respond to the Beatles' hypothesis: is love, love, love all you need? Historian Stephanie Coontz's fascinating book *Marriage, a History* deals with the difficulties human beings have always had with erotic and romantic partnership, and focuses on a conundrum: that in the contemporary Western world a new standard has been created where

> people expect marriage to satisfy more of their psychological and social needs than ever before. Marriage is supposed to be free of the coercion, violence, and gender inequalities that were tolerated in the past. Individuals want marriage to meet most of their needs for intimacy and affection and all their needs for sex. Never before in history had societies thought that such a high set of expectations about marriage was either realistic or desirable. Although many Europeans and Americans found tremendous joy in building their relationships around these values, the adoption of these unprecedented goals for marriage had unanticipated and revolutionary consequences that have since come to threaten the stability of the entire institution.[8]

Coontz's message is that the kind of perfect love we tend to imagine in today's popular culture is well-nigh impossible to find. But we search and we search for it, with the same part of our brain that wants a universal, perfect salvation; in fact, the pleasure centers in our brain that are stimulated by new romantic love are the same as those stimulated by heroin and cocaine, and most likely by certain intense mystical experiences as well. But such experiences, physiologically speaking, cannot last, and so if we do not find a way to accept that the romantic and passionate love we feel early in a relationship must ultimately transition into something cooler, slower, steadier, and a little less exciting, then we are doomed to spend life running after the constantly vanishing horizon line of "happily ever after."

This is such an easy trap to fall into, even for brilliant, well-adjusted, healthy, sane (and atheistic) individuals because of the fact that we all need loving companionship: it is not a drug from which we can learn to entirely abstain through some kind of lifelong twelve-step program. And because of the stress (not to mention wedding-planning and then couples-therapy bills) we cause ourselves in searching for it, the relatively secular myth that there is one true, perfect love for each of us, waiting to be discovered, can be at least as irrational and pernicious as many versions of God.

I'm hardly suggesting that we should not love, or not form long-term, committed pair bonds and exclusive marriages. Humanism no more encourages libertinism than it pushes monkish asceticism. In fact, performing wedding ceremonies for Humanist couples deeply committed to long-term, exclusive life partnership is one of my favorite parts of my job as a Humanist chaplain. (If you're curious: Yes, of course they should live together and have sex before marriage. It's pretty tough to choose a partner for the rest of your life without knowing if you're compatible.)

For now, however, I'm not arguing that we shouldn't look for love; merely that romantic love can't be *everything.* It cannot be the meaning of our lives.

Happiness

Happiness is the most obvious value one might choose as the goal of life. Who doesn't want to be happy? But merely wanting to be happy doesn't imply anything about respect for the happiness of others. And if I choose the pursuit of happiness as the meaning of my life, I place myself precisely in the ethical position of Ivan Karamazov, because I can claim that anything makes me happy: promiscuous sex, binge eating and drugging, murder, abuse, theft, embezzlement, conspicuous consumption, or ignorance (which is, after all, *bliss*).

Even if I add to the idea of happiness as the meaning of life the nineteenth-century American Humanist leader Robert Ingersoll's helpful qualification that "the way to be happy is to make others so," I am left with agonizing questions: Which others should I make happy? Is it enough to focus on making certain specific others, like my friends, loved ones, coreligionists, or countrymen happy, or must I treat the happiness of all people as an equal obligation? If I choose to prioritize those close to me, I might head down the road toward becoming an ethical ax murderer whose family loves him, but if I choose to treat all people's happiness as equal, I'll be paralyzed to inaction by having to combat my natural tendency to privilege those close to me and by the fact that *everyone's* happiness seems an impossible goal, with no way of working toward it that doesn't first involve either focusing on my immediate surroundings or ignoring them. Thus, if Gandhi or Martin Luther King Jr. were to be evaluated on whether they made others happy, would we laud them for pleasing the masses or criticize them for having been, by most nonhagiographic accounts, deeply flawed in their relationships with their own wives and children? This is not to say that those two leaders were failures, but rather that happiness as the standard for a meaningful life is too egocentric, too nebulous, or both.

Be All That You Can Be

Eva Goldfinger, a Humanist leader I admire, writes in her *Basic Ideas of Secular Humanistic Judaism*, "Although there is no single over-arching purpose to life, self-actualization for every human being gives life purpose . . . [Humanists] believe that the most important purpose of human life is for every individual to strive for and attain self-fulfillment—to become what each is capable of and to help others do the same."[9]

This is a popular position among Humanists and human-centered thinkers of all kinds. The psychologist Erik Erikson, for example, created a psychological model positing that human lives can be divided into several developmental stages, each involving a crisis that must be resolved successfully in order to progress toward the ultimate life goal for any person—

self-actualization. The U.S. Army created an advertising campaign with a similar, if less subtle message: "Be All You Can Be."

But political commentary on the current ventures—or misadventures—of this country's military aside, the example of a soldier's self-fulfillment does raise an important concern about this framework; as with happiness, one person's achieving his potential can be another person's nightmare. Even if we presume that those who have successfully self-actualized ought to be praised for their great self-discipline, resourcefulness, and self-knowledge, ought we not to worry that this could probably have described Joseph Stalin, Tony Soprano, or the Emperor Caligula, in Camus's play of the same name? Stalin most likely had to dig down deep within to find the emotional, intellectual, and physical strength to become the best totalitarian dictator he could be. What made *The Sopranos* one of the greatest shows on television was the way it showed just how hard Tony had to work to get good at being bad. Every day he faced ethical dilemmas: Should he choke the life out of his beloved little cousin Christopher, whose drug addiction and lousy all-around judgment have led to one too many screwups? Should he risk his men's lives by going to war with a rival gang over some petty matter of pride, or merely continue to embezzle money from and with them at a slightly reduced rate? These kinds of decisions require *leadership*—but not a shred of the kind of ethical compass of which Kant might approve. Camus focuses in on this problem with Caligula. Motivated by the death of his sister—who was also his lover—he very purposefully and successfully set out to become a diabolically murderous tyrant in order to teach his subjects various philosophical lessons, including, it seems, that they should kill him. Are these examples to follow?

Self-esteem is no better a goal in itself. Psychologist Roy Baumeister points out that one of the main causes of violence and cruelty in children is unrealistically or narcissistically high self-esteem, especially when coupled with what Baumeister calls "moral idealism." If we need to see ourselves as highly valuable and esteemed, and this comes into conflict with reality even temporarily—as of course it frequently will—young men in particular are prone to lashing out violently to defend their wounded pride. We see this not only on the battlefield but in politics and in love, or rather, in the painful, angry, embittered reactions of the injured party when he not only loses love but also takes a hit to a self-image falsely built up by praise and compliments that had nothing to do with his skills, accomplishments, or ability to help others.

Indeed, even if we focus on Goldfinger's last clause above, about helping others to self-actualize—a clause no doubt added on to avoid the kinds of objections I've just gently suggested—we run into the same problems we did with happiness. Which others should we help to self-actualize? Only those closest to us? Or should we consider ourselves obligated, like the Bodhisattva, or disciple, of Mahayana Buddhism to take on the doomed-from-the-start mission of attempting to help all sentient beings self-actualize? In the end, the entire idea of self-fulfillment amounts to little more than a psychologically dressed-up version of happiness; potentially a wonderful purpose for life, but when taken literally and seriously, potentially solipsistic and menacing as well.

Be of Service

Another popular and appealing choice as the purpose of life, whether for religious or nonreligious people, is the idea of making a difference: being of service to one's fellow human beings and working toward the welfare of humanity. Perhaps this would have been Rieux's answer? If his dedication to humanity was so evident, perhaps that is because he chose it as the very purpose of his existence, and perhaps we ought to as well. There is only one problem with this idea, which seems to provide such a potent antidote to the selfishness of Caesar and others: it may be *too* unselfish. Even if we assume that helping others is as good a thing as it seems—and we do—it simply seems too much to ask that a person choose a meaning for his or her life that does not address the idea of self-interest at all. This is precisely what communism and socialism allegedly do—they ostensibly ask each person to work for the good of the collective, to sacrifice individual desire for possessions, leisure, or even certain types of freedom—all this in the name of future generations, who will inherit a better world. But as Beauvoir chides in *The Ethics of Ambiguity*, if the individual is nothing, the society cannot be anything either. "One can not, without absurdity, indefinitely sacrifice each generation to the following one."[10]

Perhaps this is best understood in the form of a bit of folksy but wise advice that a beloved former English professor of mine, Ralph Williams, of the University of Michigan, once gave me—that you have to strive to be selfish half the time, because more often than that is no good, but if you think you're being selfish less than half the time, you're probably lying to yourself. There are countless volumes of outraged, anti-Communist fiction from behind the old Iron Curtain by writers who suffered under the brutal mandate to serve the collective under Stalinist Russia or Maoist China. These works, by authors who would certainly support the general idea of "being of service" to others, are a reminder that free people, no matter how generous, will want to choose a purpose that also affirms the value of their own individual lives.

This is far from saying that Humanists shouldn't be self-sacrificing—in fact, I agree that working toward the welfare of humanity is a crucially important part of life. Being of service to others is very important. We have all kinds of statistics to show that, in fact, people who do community service are happier than those who don't, and spouses who give more to their husbands or wives live longer, healthier lives than those who receive more.

So, spouses, let your bickering over who gets to help the other more begin. But before you get too passive-aggressive, or the martyr complex starts to heat up, remember that just as a loving parent in an airplane whose cabin pressure has destabilized must put on her own oxygen mask before attending to her children, when we explain why we ought to give and to help others, we must begin with our individual needs, and then move to others' needs, not vice versa.

Dignity

If happiness, love, self-actualization, self-sacrifice, not to mention God, materialism, and antimaterialism won't work as a purpose around which to build good lives without God, what's left? Nothing? Is there no purpose a Humanist can honestly live for?

Not only would that be a completely depressing answer, it wouldn't even pass as true or honest. We've known since Aristotle that everyone lives according to a purpose, whether consciously or unconsciously. Eliot Spitzer may have thought he was living for justice, but probably he would have been a lot better off acknowledging the extent to which he was motivated by love, sex, and probably just thrills. I meet plenty of students who reply, shocked that I've even had the temerity to ask them what the purpose of their life is, that they really don't know. But if you observe their behavior, you'll know: in the Ivy League, more often than not, people are living to achieve and to succeed. But there are also plenty who suspect that life has no higher purpose than finding a great party or high or some other form of fleeting pleasure. And look, I'm not saying I have anything against either academic or other success, *or* fleeting pleasure, but as an overall motivation for living there's something healthier available: a better way to understand what we're really after when we're at our best. And I'd love to tell you what that better way is, but there's a slight problem before I do: I'm not positive we've yet found a single word that we can all agree on to stand for this "better way." Even Camus had trouble finding one.

As *The Plague* draws to a close, the novel's narrator describes the town of Oran as it creeps back to life after months of suffering. Ships begin to transport visitors and supplies again; trains steam ahead across the chilly North African landscape; suddenly the time that moved too slowly during the dying is now passing too quickly.

The narrator turns his attention to the human relationships and connections that are reestablished, along with those forever broken. Mainly, he turns to love. Rambert and his bride are reunited: she skips down from the locomotive's steps, already running as she hits the track, throwing herself into his arms so quickly that he is crying before he can even brush aside her hair to check that hers is the face he has missed for so long: Is it really her? Or, is it really *him*? He worries that he can no longer be the man who longed, simply, only for her embrace. He has changed during the plague. Haven't we all? After all, the plague is a metaphor for what we all experience. We are all innocent victims.

In a final confession, the narrator tells us that he is, in fact, Dr. Bernard Rieux. And because the narrator is also clearly Albert Camus himself, we confirm what we have already known: that Rieux is Camus, and the doctor's tireless dedication to dying human beings in the absence of a saving God is really Camus's unwavering compassion for humanity despite all our failures.

He wrote the book, he tells us, "so that he should not be one of those who hold their peace but should bear witness in favor of those plague-stricken people; so that some memorial of the injustice and outrage done them might endure; and to state quite simply what we learn in time of pestilence: that there are more things to admire in men than to despise."[11] Camus's defiant struggle to bear witness eloquently for human potential, despite all our failings, can best be understood as the struggle for dignity. At least that's what my friend and teacher Sherwin Wine would have called it.

As I write this, I am reminded of Sherwin: he was the only person I've ever met, even after several years among the best and brightest at Harvard, who I imagine would have been the intellectual peer of Albert Camus, that fierce genius who had become one of the most important literary figures in Western history by the time of his tragic death at the age of forty-seven.

Sherwin too died in one of those ways that we call, as if all deaths weren't, "tragedies"; one of those stories that, when you hear it, makes you lose your breath and think of a hundred little memories of moments you're glad to have had with your own loved ones.

In Essouria, Morocco, two men who had loved each other and traveled the world together for thirty years had finished dinner and were returning to their hotel in a taxicab. Sherwin was in the early stages of working on a book that might have covered a number of the same topics as this one. They were as happy as such couples can be. Richard had long since gotten used to Sherwin's workaholic lifestyle and enjoyed trips like this with him. Before he left for the trip, Sherwin, at age seventy-nine, enthusiastically told his childhood friend and sometime editor Nancy that they had a "twenty-year plan" for all the books he still wanted to write and edit. That plan did not include a Moroccan man who drank too much, got in his car, and swiped into their taxi, killing himself, the taxi driver, and one of its American passengers. But such are the deaths that happen every day somewhere and make us want to say—even those of us who are not only atheists but also atheists who have consciously resolved never to pray—"There but for the grace of God go I." English does not yet have an equivalent secular idiom to convey the same sentiment.

Many of us whose lives Sherwin touched were shocked, even devastated, when we heard the news. He was such a uniquely skilled and compassionate leader that even in a Humanist movement that insists there are no saints and no supernatural authorities, it was hard at first to envision what we would do without him. But eventually we remembered that he'd been very clear about how to deal with life's tragedies: by cultivating dignity.

Wine spent many years refining his definition of this ephemeral quality, a kind of stew with equal parts love, friendship, reason, justice, and self-discipline, taken with a shot of optimism and a chaser of defiance. He defined it by describing its four qualities: "The first is high self-awareness, a heightened sense of personal identity and individual reality. The second is the willingness to assume responsibility for one's own life and to avoid surrendering that responsibility to any other person or institution. The third is a refusal to find one's identity in any possession. The fourth is the sense that one's behavior is worthy of imitation by others."[12]

Along with these four characteristics of dignity come three moral obligations for the person who values them: First, "I have a moral obligation to strive for greater mastery and control over my own life." Second, "I have a moral obligation to be reliable and trustworthy." And third, "I have a moral obligation to be generous."

For a long time, I was too intimidated to write about dignity, ironically because I saw certain people I admired, like Sherwin, who actually lived according to this definition more often than not. Sherwin was one of a number of people I knew who were not necessarily toe-tappingly happy or in love at every moment, not giving themselves away to God or to others, but never self-absorbed. They were and are strong, real. I saw these qualities in my mother, a couple of close friends, even in a few political and social leaders. Barack Obama's self-possession has this kind of dignity, and this (along with the contrast between his platform and George Bush's) was one of the first things that led me to admire him.

But I wasn't sure I saw those qualities in myself. Was I always living a life of dignity? Was my behavior always worthy of imitation? Was I always self-aware? Did I sometimes slip into identifying with some of my possessions, or with a particular achievement? Was I always as generous as I could be, helping others to conquer the plagues they encountered in their lives?

Of course not. But after months of thinking about it, and after countless hours spent getting stuck at this very point in writing this book, I finally came to a moment of clarity, sitting on a beautiful cobblestone patio, on a perfect blue day, overlooking the water at Hyannis Port, by the ferry to Nantucket. There I realized that dignity is not a state you have to get to in order to be a Humanist, and it's not even a place you need take permanent residence in to write a book about goodness without God. If it were, I could simply describe it to you and the book would end here. But I'd known I wanted to talk about dignity (or, as we'll see, whatever you choose to call it) early in this larger discussion because dignity is a goal to strive for. It's a direction to head in, a target to focus on so we know whether our aim, in living, is improving.

The struggle for dignity is a purpose that can drive us to great heights, but we would never get anywhere if we stopped in defeat, reproaching ourselves for hypocrisy at the first sign of difficulty. And so I can finally talk about dignity comfortably now, even though I freely admit I've not yet reached the point where I am a perfect example of it—and even though none of us ever will be perfect at it.

What Is Dignity?

Sherwin Wine didn't invent this concept of a meaning to life beyond God and a simplistic understanding of happiness, love, and the rest. Erich Fromm, the great twentieth-century Humanistic psychologist, called it *humanity*, or being fully human. He wrote:

> Dostoyevsky once said, "If God is dead, everything is allowed." This is, indeed, what most people believe; they differ only in that some draw the conclusion that God and the church must remain alive in order to uphold the moral order, while others accept the idea that everything is allowed, that there is no valid moral principle, that expediency is the only regulative principle in life . . . In contrast, humanistic ethics takes the position that *if man is alive he knows what is allowed*; and to be alive means to be productive, to use one's powers not for any purpose transcending man, but for oneself, to make sense of one's existence, to be human. As long as anyone believes that his ideal and purpose is outside him, that it is above

> the clouds, in the past or in the future, he will go outside himself and seek fulfillment where it can not be found. He will look for solutions and answers at every point except for the one where they can be found—in himself.[13]

Still, Fromm's language seems a little too focused on what man has in himself; we know intuitively that this is just a little bit incomplete. We *don't* have everything we need for a good life inside ourselves alone. If we did, we could all go off into separate little rooms and just *enjoy*. Rick Warren is on to something there. But rather than swinging from Fromm's Humanism all the way to Warren's evangelism, we can modify Fromm slightly with something like the approach of Jonathan Haidt, who speaks of meaning—what he calls "happiness," though Haidt's definition of the word is an extremely sophisticated one, rather like Epicurus's *eudaimonia*:

> Happiness is not something that you can find, acquire, or achieve directly. You have to get the conditions right and then wait. Some of the conditions are within you, such as coherence among the parts and levels of your personality. Other conditions require relationships to things beyond you: just as plants need sun, water, and good soil to thrive, people need love, work, and a connection to something larger. It is worth striving to get the right relationships between yourself and others, between yourself and your work, and between yourself and something larger than yourself. If you get these relationships right, a sense of meaning and purpose will emerge.[14]

But perhaps the best short description of the practice of the kind of dignity I'm speaking of can be found in a statement by Rabbi Hillel, as quoted in the second century CE Jewish text the *Mishnah*—again, our values don't have to be entirely separate from those of religion; of course some religious people have had worthwhile insights, just as "religious values" today include much Humanist thought:

> *If I am not for myself, who will be for me? If I am only for myself, what am I? And if not now, when?*

Dignity Requires Relationship

Rabbi Hillel's statement is really about where human strength comes from. There are those who think it is only a gift, given by God. They think we humans, without God, would have no strength in ourselves and that the only way we can become strong enough to bear our pain, to give love and help to others, is to pray for strength.

Others believe they have all the strength they need inside themselves. They don't need God, and they don't need anyone else either. They are entirely self-sufficient. A key moment in discovering goodness without God—in becoming a Humanist—is realizing that both these postures are crooked. We are not wicked, debased, helpless creatures waiting for a heavenly king or queen to bless us with strength, wisdom, and love. We have the potential for strength, wisdom, and love inside ourselves. But by ourselves we are not enough. We need to reach out beyond ourselves—to the world that surrounds us and sustains us, and most especially to other people.

This is dignity. It is the realization that, in one form or another, has been found in every major religion. All of us know what it feels like to realize "I am a person." But it takes a little more awareness to realize, "*You* are also a person." And it takes even greater awareness still to recognize that I am more of a person when I am helping you to be more of a person.

If I Am Not for Myself

The novel *Scar Tissue*, by Canadian politician and former human rights scholar Michael Ignatieff, tells the story of a middle-aged philosophy professor struggling to cope with his mother's Alzheimer's disease, and his worries that he too will one day be seriously ill. The protagonist's brother, a doctor, introduces him to a patient, Moe, who has amyotrophic lateral sclerosis, better known as Stephen Hawking's disease. Moe's is a "shut-in" existence where he has no more control over his body than the ability to manipulate a breathing tube that allows his thoughts to be typed out on a computer screen. Yet he enjoys life. His thoughts, while physically painful and further damaging to his nervous system to express, are eloquent and inspiring. The professor is forced to consider that the human will can overcome a great deal.

But what about when we literally have a hard time being self-aware? Moe's dignity in the face of debilitating illness enables the professor to realize that his mother, who is also ill, still possesses dignity—albeit of a slightly different sort. The mother, though no longer able to communicate verbally because of her Alzheimer's disease, gives her son a number of signs that, despite her illness, she will never, until the very end, stop struggling to recognize and relate to him and to herself. Her courage is not only inspiring to her son, it is *valuable*—it enables her to be a moral role model for him, giving him a sense that it would be possible for him too to die from Alzheimer's with dignity, in the face of his fears that he would inherit her illness. By living for themselves in spite of difficult circumstances, such courageous people are of more help to others than even they realize.

If I Am Only for Myself: Other-Awareness

In Tolstoy's novella *The Death of Ivan Ilych,* Ivan is a man who will do whatever he needs to do to follow fashion and status. He takes the right job, marries the right woman, makes himself a big success, but never really sees anything or anyone except for what they can do for him. This goes on for years, but eventually he gets a mysterious, undiagnosed illness and becomes convinced he is dying. Face-to-face with his mortality, all he really wants is to be fawned over, adored, and pitied like a small child. He is aware only of his own needs, and it becomes clear that, at least metaphorically, this more than any specific ailment is what causes him such an agonizing death. He screams in excruciating pain for the last several minutes before his final expiration.

Ivan Ilych is a warning about the dire consequences of not learning how to care for others and generously help them. Because if we can't really care about other people's pain, then we can't connect with them. And if we can't love others, we can't feel close to them, and therefore we can end up totally alone—even though physically we might be surrounded by dozens, even hundreds of so-called friends. How is it, after all, that we can feel alone when we're in a city of a million people or ten million people, or for that matter in a world of eight billion people? It hardly makes sense. But of course aloneness is not a matter of whether you're around people or not; aloneness is a function of whether you can love them and allow yourself to be loved by them. And you can't do that unless you can put yourself in their heads and allow yourself to feel some of their pain when they bleed.

In Shakespeare's *The Merchant of Venice*, the tormented Shylock asks, on behalf of his fellow Jews—and indeed on behalf of all those who have been physically or psychologically victimized by those who rationalized their brutality by viewing their victims as a less human "other"—"If you prick us, do we not bleed? . . ." Any normal human being knows: when you prick me I bleed, when you hurt me I hurt. We know what it means to feel our own pain—and our own pleasure. That's self-awareness. Without it, you're nothing. But while any two-year-old knows that when he is pricked, he bleeds, it takes him much longer to fully understand that other people feel the same pleasure and pain, in the same ways. And it is only the most frightening of creatures who never learn this. In fact, what makes Mary Shelley's *Frankenstein* such a classic story, far more important than all the cheap, kitchy Halloween decorations would suggest, is that in some ways, Frankenstein's monster understands both his own pain and that of others better than Victor Frankenstein ever learns to. The monster, who understands that normal people cannot love him because he is "so very ugly," merely longs for companionship. If he can offer his love to one single understanding soul, he says, he will spare the world the great anger that is the dark side of that love. But the pathetic Dr. Frankenstein never learns how to treat his creation as a truly living being until it is far too late—and though he never fully learns to weep for the pain he has caused—only for his own pain—the monster does weep over Victor Frankenstein's grave. "He was my father," the monster says, letting go of his anger—and demonstrating the humanity—the dignity—that his creator could not.[15]

Let's hope this is a lesson that is not lost on some of the potential suicide bombers Tom Friedman described in a *New York Times* column, people who are tempted by the Western world, but believe that the Allah of radical Islam is God 3.0 (with Chrisianity as 2.0, Judaism as 1.0, and Hinduism as 0.0. I suppose this would put atheism or Humanism well into the negative numbers). Such young men and women, he writes, suffer not from "the poverty of money, but the poverty of dignity and the rage it can trigger."[16]

I'm not sure how Friedman meant the word *dignity*, but it fits precisely with the definition we're using here: the suicide bomber experiences a corruption of both sides of dignity. He is tempted by the Western world—and lacks a sense of self—but believes his religion is the best—and therefore lacks a sense of empathy for others beyond his group. This makes him (or her) angry and lonely enough to literally explode in rage.

Definition Problems

Perhaps you're still not convinced that the word *dignity* is the best way to describe the "meaning of life," even for an atheist or agnostic. After all, the idea that one word can suffice for the meaning of our lives *should* be a little suspect.

A bit of background about the word *dignity*: It has ancient roots tracing back to the Latin *dignitas*. Like many words in use by philosophers and theologians today, it used to mean some things that it does not mean anymore in this context. Dignity used to mean a sense of being externally valued by others, for example: your dignity was how much you could be sold for. The philosopher Hobbes said that in times of war a soldier has more dignity, whereas in times of peace, a politician has more dignity, because dignity is how much a person is worth in any given situation. This idea obviously sounds repugnant to us today.

But there's another definition of dignity in use today by the Charter of the United Nations: "We, the peoples of the United Nations, determined to save succeeding generations from the scourge of war, which twice in our lifetime has brought untold sorrow to mankind, and to reaffirm faith in fundamental human rights, in the dignity and worth of the human person, in the equal rights of men and women and of nations large and small . . ."

This sort of dignity is "inherent" from birth in "all members of the human family," as expressed in the 1948 United Nations Universal Declaration of Human Rights. This is a much more appealing vision than Hobbes's. But it raises the question: if dignity is inherent in all people, then what does it mean to earn it, or to live by it? Why should it be such a great accomplishment to live according to something you were born with, anyhow?

The existentialists also run into this problem. Sartre talks about freedom as something we all have and must come to terms with—a basic, inescapable fact of life: "man is free, man *is* freedom."[17] But *promoting* the kind of inevitable quality Sartre is talking about would be sort of like promoting the aging process, or marketing ice to an Eskimo. Meanwhile, Sartre's lover, the philosopher Simone de Beauvoir, talks about freedom as the key thing all human beings must try to achieve, for themselves and for all others: "The cause of freedom is not that of others more than it is mine: it is universally human . . . If I want the slave to become conscious of his servitude, it is both in order not to be a tyrant myself . . . and in order that new possibilities might be opened to the liberated slave and through him to all men."[18] These sorts of semantic differences must have made for some interesting pillow talk.

Perhaps Fromm solves the semantic problem, arguing that a free person is the tree that comes from the seed of freedom we are all given at birth, and that dignity as a personal achievement is a flower that emerges from the seed, the potential for dignity that each person possesses:

> If we say that the tree is potentially present in the seed it does not mean that a tree *must* develop from every seed. The actualization of a potentiality depends on the presence of certain conditions which are, in the case of the seed, for instance, proper soil, water, and sunlight. In fact the concept of potentiality has no meaning except in connection with the specific conditions required for its actualization. The statement that a tree is potentially present in the seed must be specified to mean that a tree will grow from the seed *provided* that the seed is placed in the specific conditions necessary for its growth.[19]

For human beings who have the potential for dignity, then, they will attain it *provided* some luck, health, love, nurturing, and hard work on themselves.

Perhaps you still can't get excited about the word *dignity* because, like Steven Pinker, you're concerned about the way it is sometimes manipulated by religious scholars to serve reactionary ends. In an e-mail sent after returning from a frustrating forum on bioethics held at the White House by advisers to President George W. Bush, Pinker wrote, "I learned today that the intrusion of religion into government forums on bioethics is very much a Catholic phenomenon, especially the emphasis on 'dignity,' a meaningless word that is being used to convert Catholic doctrines into restrictions on biological research. The Protestants really are more concerned these days with 'justice'; they can't get excited about 'dignity.'"[20]

My point is simply to identify a concept—and the word *dignity* here is less important than the concept itself, so let's not get hung up on the word. Maybe this concept of a deep, rich purpose that can drive the lives of millions of people to be good without God is something for which we've never had a perfect English word. So call it "integrity" if you like, or "flourishing," or "*humanity*," or call it "x" if you like. But there is a state in which you're aware of your own humanity, and you're also aware of others' humanity, and you're aware that all human beings are human. There's a state in which you're aware of your own vulnerability and mortality, and that awareness allows you to connect with others from a place of strength and empowerment. There's a state in which you don't have too much clingy connection or too much lonely disconnection, but where you combine self and other. Being in this state feels good in both the short term and long term—good enough to motivate us strongly. And so our goal is to get there and try to stay there.

For me, writing this book has been a profound test, and ultimately a confirmation of my belief that dignity is a worthwhile purpose. There came a point in writing where I just wanted to give up. It felt too hard. It's not that I didn't believe in the message. I didn't believe in myself as the messenger. Sometimes I felt as if I was too young to tell the story of Humanism. There were so many older people out there with more experience, more knowledge. Then I would remind myself that Humanism is a tradition that traces its roots back well over two thousand years, across many continents. It encompasses the best of our scientific, historical, literary, and philosophical thinking over that period. So no one, regardless of age, should ever feel perfectly confident in her or his ability to tell this story of goodness without God. I have the training, and I've been thinking about these issues all my life. Why not me?

But mainly I was just afraid to fail. A few months into the writing process, I spent a few days with my father's first cousin, Ruth Seymour, the venerated—and intimidating—director of KCRW, the Los Angeles NPR affiliate. Ruth has been such a towering figure in public radio over the past generation that some people in the industry have called it "National Public Ruth." Ruth and my father were close when they were kids. They were the same age, and similarly gifted. But my father wasn't a success the way Ruth was. He lost his way somewhere—no one is exactly sure where—and never fully made it to a satisfying career in either of the areas that interested him most: writing and religious mysticism. Ruth and I sat and talked about what I'm doing with my life, about Humanism, about why I wanted to write this book.

After hearing me go on for a couple of days, she said something that struck me hard: my father would have given his right arm for the opportunity to do what I'm doing. It sounded true, maybe even literally. After all, with the lung cancer, he actually gave his right lung just to stay alive a few more years. So why not his right arm to find some of the meaning and purpose he never quite found? I cried for my father that night—for the pain he must have felt, not only battling his illness but, as far as I understood it, battling the ravages of the disease without any clear sense of a mission in life to make his struggle more worthwhile.

And then I cried for myself. I cried because despite having attained many things by thirty-one that my father never achieved in his fifty-nine years—I had a fascinating career, a set of beliefs I was sure of, a chance to make a difference, and a book contract—I was still afraid to fail. I was worried about what people would think of me, and what they'd think of my book. I just thought the task of composing this volume was too big and too tough and I couldn't do it. I was already ashamed of myself for the failure I anticipated. I began to wish I'd never tried my hand at writing at all. I wasn't thinking of harming myself, or of trying another career—I was just having trouble getting out of bed in the morning. I'd made myself depressed.

Still, I fought through it, working what some people might consider hard, though I knew in my heart I was capable of working harder. All summer I wrote, and I was proud of the way I expressed some ideas about Humanism. But it took intense effort, every hour I sat down, just to get myself to face the blank screen of my computer. In my heart I still wanted to quit, still thought I was going to fail. I was afraid to let my father down, yet subconsciously afraid to step out and succeed where he had failed. Maybe if he hadn't done it, I wasn't meant to. Finally, at the end of the summer, I was getting nervous about my deadline and I decided to spend a week out on Cape Cod at a little bed-and-breakfast, doing nothing but writing for six days.

The week went pretty well, but it passed quickly. By the last night I'd made progress, but not as much as I'd hoped to make. That night I went out for a nice dinner with my hosts, then came back and was feeling predictably tired after a big meal of good New England seafood, so I switched on the television instead of getting back to work. The first thing I flicked to was a movie about high school football—one I'd never heard of, called *Facing the Giants*. I started watching: just what I was in the mood for that night, because the characters felt even more down on their luck than I did. The high school football coach was a lovable loser. His team stunk and the parents wanted to fire him. His adoring wife desperately wanted to get pregnant but his sperm count was too low. His car wouldn't start, and there was some kind of foul smell in his kitchen. In short, he was a redneck Raskolnikov. And his players were no better—a collection of misfits, malcontents, and scrawny freshmen.

Now I know sports movies, and I could smell the team's heartwarming if unrealistic transformation into champions coming from a mile away. But it turns out I'd happened upon not only a *Christian* transformation story, but an Evangelical film inspired by none other than Rick Warren's *A Purpose Driven Life*. The coach not only found himself just in time to save his job, motivate his team to unforeseen new heights, get his wife pregnant, and even find and remove the dead mouse that was making his house stink—he did it all by realizing that it wasn't "about him." He'd been too worried about his own wants and not thinking about what God wanted. He'd been coaching football under the tacit assumption that a high school team's mission was to win games, not realizing that the real purpose of the team was to be a way for the players and coaches to give glory to God.

Well, maybe it's that all this schmaltz was actually conveyed pretty decently by the filmmakers, or maybe it's just that I'm a sucker for heartwarming movies, or both—but I cried. Jonathan Haidt and some of his colleagues at the University of Virginia have studied an emotion they call "elevation," the feeling you get when you hear an inspiring story of personal triumph against all odds. Among other things, Haidt's team discovered that heightened oxytocin levels even cause nursing women to lactate more when they're feeling elevation. And elevation is what I was feeling watching the film, seeing this depiction of a Christian message transforming the characters' lives, these lovable losers learning to conquer their fear by resolving to give their very best possible effort, no matter how much it hurt, or how afraid they were of failure, and trust that if they served God, he would give them whatever they needed. I guess they needed the state championship, because that's what they got.

When the film ended, I felt a rush of recognition and a renewed sense of responsibility as a Humanist and as a writer about Humanism. I knew how many young people out there were watching movies like that, using Warren's message to help them conquer their own insecurities. In order for Humanism to have the impact I believe it ought to have, it would not only have to be rational and scientific and *true*, it would have to have a message that could inspire people to overcome their fears. And I knew damn well there was no way I could succeed at helping others find courage if self-doubt was still hanging like a fifty-pound weight on every sentence I wrote.

I got up from the couch and saw my own tearful eyes in the mirror. I thought, "I may not have an all-powerful God to turn to, but I've got to find some way to beat this malaise. What is it?" At that moment, I actually asked myself, what would happen if I just *pretended* to pray to God for guidance, just to get the same psychological effects—the placebo effect that would get me into a state of mind where I'd be able to do my best work? I considered the idea for a few seconds before I realized how stupid it was. I couldn't use *make-believe* to help me write a book about Humanism's empowering message of truth.

Next I asked myself why I was having so much trouble finding the right words to offer, and finding so much inspiration in this Christian message, when I was so convinced that Christianity's supernatural story of a resurrected God, born of a virgin, dying for our sins, was simply a human-created myth.

Thankfully I then finally realized what attentive readers most likely figured out many paragraphs ago. That Rick Warren was right about the way I was approaching my work. He wasn't right about Jesus, or God, or heaven, or hell. He was dead wrong on gay marriage, abortion, and about atheists being more arrogant than Christians. But he was absolutely right in the sense that I'd fallen into the trap of thinking that my work, in this case writing this book, was all about me. I'd spent too much time worrying about what I was going to say, or what readers would think of me, and not enough time thinking about how Humanism could truly be helpful to others, and how my book could help people to attain the knowledge and help they wanted. Don't get me wrong—it's hardly true that up to this revelation, I'd been self-interested 100 percent of the time, or that I hadn't written anything with a view to helping my readers. Of course I had. But I'd developed a pattern wherein when the work got tough and feelings of fatigue would start to set in, I'd immediately start to think of myself as a failure and worry that I was going to let people down, rather than training my mind to simply do my best, think of the Humanist message and of what others needed to hear, and keep going.

Suddenly, once I really acknowledged that the purpose of this book was less to advance my own career and more to help others see Humanism as a tool to advance their lives and address their fears, my own fear of failure began to melt away into insignificance. Because I understood that if everyone fears failure sometimes, my own fears just weren't that big of a deal. What was more important was what I did about my fears, not only for my own benefit but also to benefit my loved ones, other people, and the world I'm fortunate enough to live in.

I realized that finding courage and discovering dignity is kind of like walking a tightrope—not that I've ever been in the circus, but I've done a few high ropes courses in my day. On a tightrope, if you spend your time looking down and thinking you're going to fall flat on your face, you probably will. If you're constantly glancing ahead too far, thinking about how cool you're going to look once you've made it to the other side, you're going to eat plenty of air, and maybe some dirt too. But if you simply practice hard and keep focused on moving forward one step at a time, at the very least you're going to have a good trip. Similarly, we're bound to fall on our faces anytime we allow ourselves to become obsessed solely with our own needs, or anytime we give up entirely and allow others to walk all over us. We trip and fall anytime we get so insecure that our own fear prevents us from recognizing and reaching out beyond the fear of others—or anytime we get so arrogant that we can't allow ourselves to learn a profound lesson from a person or group with whom we often disagree.

If we practice keeping Hillel's advice in mind from moment to moment, bringing our attention back to his message every time we inevitably get distracted—if we learn to balance our self-awareness with generosity, and our self-possession with a concern for others in the world beyond us, we'll do well. We may not achieve every goal we set for ourselves. Heck, our football team might not win the state championship, and our books might not be the biggest best-sellers since the Bible. We probably won't even feel *happy* 100 percent of the time, because presumably we'll still be living on earth. On earth terrible things sometimes happen and there's no reason to smile about them. But committing ourselves to get involved in worthy, human, this-worldly causes that are truly beyond ourselves will bring with it a feeling—call it dignity, integrity, grace, freedom, or whatever you like—which will be the same feeling Rieux had, that kept him motivated to help all those plague victims without needing any hope of a reward in heaven. It will be the same feeling all the real-life secular Rieuxs of Doctors Without Borders get when they're out on a difficult, thankless, dangerous mission in Darfur or Haiti. It will be the same feeling Sherwin Wine felt when he stood tall and worked tirelessly in the face of withering criticism, to build and serve a congregation and a movement of Humanists. The dignity of mutual concern and connection and of self-fulfillment through service to humanity's highest ideals is more than enough reason to be good without God.

CHAPTER FOUR

Good Without God: A How-To Guide to the Ethics of Humanism

What is a secular, Humanist, modern morality? The question becomes particularly pressing, and its answer increasingly uncertain, as more and more secular people come to fear that our contemporary conditions are precisely what alienate us from our true and better nature. We tell ourselves we're so stressed because of all the smog, neon advertising, greasy calories, bad TV shows, and Internet pornography. We like to imagine that modern life, including the demystification and desacralization of the universe by means of science and secularism, is why we have too little respect for our elders, too much casual sex—in short, the popular narrative goes, modernity is precisely why morality is falling apart. It's why good lives are falling apart. It's why everything is falling apart.

But the truth is, we've been dealing with questions of morality and of stress, alienation, and shock for a lot longer than we've had *modernity* in our vocabulary. There are so many ancient Greek and Near Eastern quotations about "today's youth" being degenerate and immoral that scholars have actually lost track of who originally said them. But the first known person to systematically diagnose the real problem might have lived twenty-six or so centuries ago, in India. For the Buddha, humanity's problems certainly weren't modernity, or even difficult external conditions. The problem was human mortality itself. And so he pioneered a moral system intended not to address any particular social ill so much as to focus directly on the three most basic difficulties every human being faces: sickness, aging, and death. Here's my adapted version of his story.

The young man's name was Siddhartha. He was a local prince. He was born to a wealthy and powerful father in a village in India, around two and a half millennia ago. The climate in Siddhartha's birthplace was sunny and pleasant, and food supplies were ample and rich. The women, healthy and beautifully groomed and perfumed, were solicitous of him because of his station, good looks, and natural charm. There were many servants to help with his every need. And Siddhartha's doting father spared no expense to ensure that the prince would encounter none of the struggles his father had endured in earning all of this prosperity.

In short, life was perfect—and therefore, by the time he reached young adulthood, Siddhartha was bored out of his mind. Without any conflict, alas, human life is excruciatingly tedious.

One day, in order to break the tedium, Siddhartha gathered up a few servants and supplies for his first-ever unsupervised journey outside the family compound and the village gates. He slipped away against his father's commands. They had been so afraid he would suffer out there, alone in the world. But on his departure he was filled with anticipation, ready to discover the untold joy and beauty that had been denied him.

Not long after setting out, they came across a woman experiencing some sort of condition with which Siddhartha was not familiar. Her skin was pale and blotched. Her brow was drenched in sweat, but she shivered as though freezing. She was calling out softly, noises that resembled the "Om" of Siddhartha's daily meditations, but with an expression of nervous inattention and in a tone of great weakness and unhappiness that was nothing like the Om's clarity: "Ohhhhh . . . Ohhhoh . . ." Siddhartha asked after her and was told the woman was sick. "*Sick?*" Yes, they told him. How could he not have known? Sooner or later everyone experiences sickness.

This worried Siddhartha a little as he began to feel a sympathetic pain for her, a pain he had never before known. But they continued on, as there was still so much to see.

Next they came across a couple of men in another physical state with which he was entirely unfamiliar. Their skin was wrinkled and yellowing, their hair white and mostly gone, their teeth rotted, and their bodies gaunt and frail. They were having great difficulty moving along the road on which Siddhartha's small caravan met them. "What is this?" he asked, and was told that these men were old. "Old?" He was familiar with the concept in the abstract: his father and mother were older than he, and there were some physical differences between them as a result. But Siddhartha's parents had been careful to keep him away from anyone profoundly aged, so he would not have to feel . . . what he was feeling now, the worry that he too would one day look and feel that way. The possibility of experiencing something so different was at first fascinating, but he had been taught to be proud of his supple skin and hair, his strength and athleticism. So the notion of aging as the permanent loss of such things suddenly became a burden on his heart, and he asked to be taken away from the old men, quickly.

Finally the group of master and servants came upon the most novel sight yet. Lying farther up the road were a man and woman who seemed to be sleeping the most troubled sleep. Their faces looked pained the way the sick woman's had, and their skin was ashen as the old men's had been. There were many flies and other insects dining on open sores on their bodies, yet they did not so much as move to swat the creatures away. And worst of all, they gave off a horrid smell that made it difficult for Siddhartha to even approach and see what could be wrong with them. The travelers were dead, he was told. *"Dead?"* Yes, this meant that they would never wake up, never feel better, never again see their parents or taste a mango or anything at all. He was shocked.

What to do about all of this? Not long afterward, he sat and began to reflect. This new knowledge of human illness, aging, and death demanded a response. His life would never be the same now that he understood that he himself, along with everyone he loved, was vulnerable to such a frightening fate.

This sheltered young man eventually came to be known as the Buddha, and the strange new sights he saw on his way were painful enough to shake the foundations not only of his own life, but ultimately of an entire civilization.

I've taken some liberties here with the details of the twenty-five-hundred-year-old foundational story of Buddhism, in order to condense it—feeling free to do so because the Buddha's story is almost completely lacking in reliable historical documentation from a time even close to when the Buddha is said to have lived. Like Jesus's, the Buddha's life may be more myth than historical fact. But *myth* does not always need to be a dirty word for the nonreligious: some ancient stories have been preserved not because of foolishness or wickedness but because they contain real insight, worthwhile wisdom.

The Buddha's story is an archetypal account of the most basic human dilemma: our mortality. We must confront the reality that we and everyone around us will (if we're lucky) age, sicken, and die. Sooner or later we all come face-to-face with bitter realities: injustice, heartbreak, back hair. Life, in short, is a minefield of pain, and no one escapes unscathed. It can be a terrifying prospect. The ethical system generally attributed to the Buddha, however, is but one way of responding to life's inevitable challenges. It centers on renunciation, on taming desire and transcending this world.

Humanism's basic focus is different—it is about engaging with life, acknowledging the reality of aging, sickness, death, and other problems so that we can learn to most fully appreciate the time, health, and life we have.

In the last chapter we dealt with the question of why, without a God, we ought to feel motivated to trudge through this often frightening reality toward a good, ethical life. We explored different ways to motivate ourselves not to succumb to fear and nihilism. And we determined that the best motivation Humanists have, our understanding of the "meaning of life," is the attempt to achieve a certain mental state—call it dignity, humanity, flourishing, freedom, rebellion, or whatever. In this state we feel a sense of integrity, a security in the knowledge that we have done our best not only for ourselves but for others. When in this state toward which we strive, we have been generous enough to know that we are strong.

Achieving a sense of meaning and purpose is an important insight. But it is not the sort of thing where a writer such as I can simply inform you, "The meaning of your life can be found in working to achieve dignity," and you will suddenly gain Enlightenment, your problems solved. Because even if you have a goal to strive for—like wanting to achieve dignity—you are still left with the question, as wide open as the Grand Canyon—how to do it?

How do we Humanists and nonreligious people respond when confronted with the kinds of problems the Buddha faced? What does it mean to achieve dignity when you are dying of cancer? Is it possible to be "flourishing" in a children's refugee camp in a third-world war zone? How exactly does one go about achieving "freedom" when one has a limited amount of time and strength and the world is filled with so much injustice? In short, we now know that we *can* be good without God, and we know *why* we want to be, but we must still answer the question of *how* we should go about being good without God, or *what it means*, concretely, to be good without God.

What Is Ethically Good?

There is no good except between people. We can't speak of a good planet, sun, or moon. There are not "good" frogs or "bad" butterflies, except as these creatures affect us humans. There are no "bad" lampposts or "good" motorcycles, except insofar as they give us pleasure or pain, benefit or harm. A good medicine is one that cures people. A good environmental policy is one that renders the earth inhabitable for human beings and other living creatures. And a good God, in theory, would have to be one that did good things for people too. Could we even intellectually conceive of calling a deity *good* unless it behaved and bequeathed in such a way as to benefit humanity—or at the very least whatever chosen group of humans we happened to be a part of? Anything else would be absurd even for religious people. Yes, some intellectually sophisticated thinkers do speak about how God works in mysterious ways, how he has to allow freedom, but all those explanations ultimately lead back to how God's *overall, long-term* plan has our best interests at heart. We'll get to heaven, these religious explanations assure us—the good will eventually be rewarded and the wicked will eventually be punished.

In Christianity this is obviously the case. Most people, whether or not they've read their Dante, or even the CliffsNotes to his *Divine Comedy,* are well aware that Christianity promises paradise, purgatory, and/or inferno as appropriate, to make up for its seeming inability to ensure that people receive their appropriate rewards or punishments in *this life*. And who can fail to be aware that at least certain very influential strains of Islam place great importance on the concept of heaven as an eternal reward for those whose earthly fate is less than satisfactory? We've all been exposed to too many grim visions of suicide bombers blessing themselves, off to a final bounty of seventy-two virgins, or seventy-two white raisins, or whatever it is they receive.[1] (Has it been asked, by the way, whether female suicide bombers are "rewarded" with seventy-two male virgins in the afterlife? How cruel: the ultimate sacrifice, only to be rewarded with an eternity of babysitting!)

But most people in the Western world are less aware, or not aware at all, that essentially all the world's major religions were founded on the principle that divine beings or forces can promise a level of justice in a supernatural realm that cannot be perceived in this natural one. Most modern American Jews, for example, will swear up and down that Judaism is not concerned with such things—that in fact Judaism doesn't even care *what* you believe—that it's about what you do. I sympathize with whomever I hear reciting this contemporary party line: after all, they've been told so repeatedly by their modern rabbis—rabbis who know well, whether consciously or unconsciously, that if they were to focus their sermons on the resurrection of the dead—a central concept in Jewish theology for thousands of years—they would be out of a job, because today's congregations don't want to hear about it.

Yes, whether or not this is included in the often Madison Avenue–level marketing plans put together by this generation's synagogues to attract young Jews, eternal rewards and punishments are a key part of traditional Judaism. The idea traces itself back at least to the Biblical character Job. Job, a righteous man who supposedly obeyed God, was the subject of a bet between God and Satan. (Again, how often do liberal Jews or Christians talk about things like *bets between God and Satan*? We are very aware that many Evangelical and other conservative Christians believe in and think a great deal about the Devil, but because liberal clergy have learned to politely ignore the subject, we forget that it is a key part of the founding tradition of *all* branches of Judaism and Christianity.) Anyhow, God is proud of Job's devotion to him, but Satan devilishly insists that Job only loves God because he has prosperity on earth: a good family, a livelihood, etc. Job will curse God just like all the other impious mortals, Satan insists, once he experiences greater suffering. And so the bet is launched, and Job becomes a pawn in a high-stakes theological game.

Of course, after his wife and children are murdered, and after he is forced to suffer intense physical torture, Job does eventually give up the ghost of his piety, questioning God's goodness. And while God initially responds with an angry, hectoring sermon in a whirlwind, he does in the end try to show his benevolence by remunerating Job for his losses—though strangely, Job receives not his own wife and children restored to life, but a *new* family, which you'd think must at the very least have been a little awkward for him. (Note: Biblical scholars are split on whether this somewhat happy ending was actually in the original version of the story or was stapled onto it by later Biblical editors to take the edge off its theological message. Some scholars who think the ending was originally less "happy" speculate that the book may originally have been intended as a kind of secular literary parable, a skeptical—perhaps even proto-Humanistic Jewish—*critique* of the idea that God is all-loving and all-just.) And lest one think that this theme is somehow limited to the Bible or to ancient Judaism, the concept of "*Techiat ha metim*," Hebrew for resurrection of the dead, is a central part of the prayers that all Orthodox and conservative and most Reform Jews recite every single time they pray.

The world of Eastern religion, meanwhile, is no less prone to the idea that there must be a promise of good and bad things appropriately meted out in a future life to make up for the inappropriate or irrational distribution of such things in this life. What is Nirvana, after all, if not a promise that immortal perfection can be attained by those who merit it? Moreover, "Pure Land" Buddhism, the largest movement in Chinese Buddhism until it was largely wiped out by the Communists, is built around promises of magical salvation in paradise after death, available only to those who have faith in the Buddha and say special prayers to him. In fact, the most popular religious practice in all of China up to the modern era was simply reciting "*A-mi-to-fo*"—Amitabha, the Buddha's name—sometimes in ecstatic, ejaculatory tones—in the hopes of being freed from this world of pain and suffering and transported to a "Pure Land" not altogether unlike the popular understanding of a Christian heaven.

Other Buddhist sects have their own conceptions of heavens and hells, bad people who die and are reincarnated as "hungry ghosts," and good people who become like gods in their next lives. But you won't usually hear about all this in the current Dalai Lama's speeches and books, or in an article for O, *The Oprah Magazine* by some Western-born Buddhist guru. It may be a historically important part of Buddhism, but it's not necessarily what the American target market is looking to hear these days.

Hinduism famously includes the notions of Karma and reincarnation, which also brilliantly provide a supernatural incentive structure not unlike heaven and hell. The idea that you can be reborn as anything from a privileged Brahmin to a lowly Untouchable depending on your conduct is really about saying that while distribution of status may seem arbitrary in this life—some Brahmins are nasty people and do not deserve the good fortune they were born into, whereas few Untouchables deserve their oppression—after this life, there will be other opportunities for the universe to give everyone his or her just deserts.

Of course, there are liberal versions of each of these religions that do not emphasize heaven and hell. But again, these are either strains that were anything but mainstream in the past and have been given more prominence today because they are more appealing to the modern believer (such as the prominent use of Spinoza's definition of God by many whose ancestors would have damned or even exiled those who put such a liberal notion forth only decades ago) or they are modern inventions that claim more connection to the ancient tradition than is really historically justifiable (such as the Episcopal theology of a Bishop John Shelby Spong, author of *Jesus for the Nonreligious*, which ideologically owes far more to Humanism than it does to Saints Augustine or Aquinas).

The point of going through all these various traditions and calling attention to their various assertions about heavens and hells is that any tradition that puts forward an idea about what is "good" is ultimately talking about what is good for human beings. Try as we might to come up with another, seemingly less human-centered definition of the word, even within traditional systems, all things we call good really do come back to us in the end. And that's okay—just because we're concerned about making the world a good place for human beings to live in doesn't mean we are self-centered or that we don't care about anything but humanity.

So the question then is simply, which concept of what is good for humans do you prefer? The traditionally religious notion that good is what a God or supernatural force wants and commands? Or a secular system that enables us to evaluate our own needs and decide how best to try to fulfill them? Recalling that not all secular worldviews are Humanistic, below is a primer on how Humanists conceive of and attempt to uphold their vision of the good.

The Golden Rule

There is a story about an Ivy League philosophy professor who was sitting and having a coffee on a stairway in a busy street in New York City, near where a big political protest was taking place. A policeman walked over and informed the man that he wasn't allowed to sit on those steps—that he was blocking the flow of traffic in and out of the building. The professor turned his head to look from side to side and raised his eyebrows quizzically in one exaggerated motion, as if to say, "I don't see anyone here to block!" The cop, annoyed by this reaction, pointed to the protest nearby and asked, "What if everyone did what you're doing?" The nonplussed professor muttered under his breath, "Who are you, *Kant?*"

A couple of hours of jail time later, the professor finally succeeded in explaining that he wasn't calling the officer a nasty name, but rather referring to the philosopher Emanuel Kant and his most famous idea, the categorical imperative: the idea that actions can only be considered moral if they could be imitated by anyone else and produce good results.

If Kant's categorical imperative sounds somewhat familiar even to those who've never quite made it though every jot and tittle of his page-turner *The Metaphysics of Morals*, it's probably because the notion is so similar to another shot-glass-sized concept of how to be good that shows up all over the map of human intellectual history, especially in religions—often referred to as the "golden rule."

For many self-respecting secular intellectuals, the idea of a golden rule is enough to make one nervous—maybe even a little twitchy. It might conjure up images of crotchety Sunday school lectures, or TV commercials about Mormons with clip-on ties and short-sleeved dress shirts. More relevantly, it might smack of moralizing, and as my friend Scott Brewer, a philosophy professor at Harvard Law School, likes to say: "Where moralism goes, hypocrisy will surely follow." (He admits he's got a long way to go to catch Nietzsche in the pithy aphorism department, but nonetheless his point is well taken.) The well-justified allergy we have to hypocrisy is the reason George Bernard Shaw said, "The golden rule is that there are no golden rules."

Healthy skepticism aside, though, there is a concept of how to be good that may be worthy of the nickname "golden," because it really does show up again and again in basically every religion. As Lloyd and Mary Morain point out in a book called *Humanism as the Next Step*,

> Throughout the ages religions of many kinds have contained a common spirit. We can see this in parts of their scriptures.
>
> In Brahmanism we find: "This is the sum of duty: Do naught unto others which would cause you pain if done to you" (*Mahabharata*, 5, 1517).
>
> In Buddhism: "Hurt not others in ways that you yourself would find hurtful" (*Udana-Varga* 5, 18).
>
> In Christianity: "All things whatsoever ye would that man should do to you, do ye even so to them: for this is the Law and the Prophets" (*Matthew* 7, 12).
>
> In Confucianism: "Is there one maxim which ought to be acted upon throughout one's whole life? Surely it is the maxim of loving-kindness: Do not unto others what you would not have them do unto you" (*Analects* 15, 23).
>
> In Islam: "No one of you is a believer until he desires for his brother that which he desires for himself" (*Sunnah*).
>
> In Judaism: "What is hateful to you, do not to your fellowman. That is the entire Law; all the rest is commentary" (*Talmud, Shabbat* 31d).
>
> In Taoism: "Regard your neighbor's gain as your own gain, and your neighbor's loss as your own loss" (*T'ai Shang Kan Ying P'ien*).
>
> In Jain scriptures: "The essence of right conduct is not to injure anyone."[2]
>
> But varying religious practices and diverse theological beliefs have been built upon and allied to this common ethical basis.

The point is obvious but achingly, embarrassingly important: the very first thing we have to do in order to be a good person is learn to look inside ourselves, understand what we love and hate, and use this information when deciding how to treat others. I say it is achingly important because it hurts to think about how often people brush us aside, ignore us, or get angry or wrong us because they are thinking of us only as little pink and brown objects in their way, not as human beings who will feel the same way about their behavior as they would if they had to endure it. And the golden rule is embarrassingly important because it is humiliating to think about how often we ourselves often buzz right past our kids or our spouse or our best friends, eyes distracted, focused on some goal or fantasy we have about how our day ought to be going, forgetting that these people too are struggling not only with petty everyday problems but with their own fears about aging, sickness, and death. Our dignity begins to slip away when we lose sight of our ability to stop and acknowledge their existence, and their struggles, for a moment.

The golden rule shows up in every religion because, for the reasons we discussed in chapter 1, religion has shown up playing a prominent role in just about every society. You can have a society that doesn't have Krishna, Jesus, or Buddha and it will be fine. Eliminate multiple prayer sessions per day, gift-giving around the winter solstice, or candle-lighting every Friday night, and things will work themselves out.

But if you have a society that lacks this principle? Then all hell really will break loose. Then you don't have a society. You have chaos.

In the killing fields of Cambodia and Rwanda during their genocides, religion was not absent, but the golden rule was as hard to find as a respite from death. People were thinking only of their own pain and their own wants, and the pain and wants of others—*the lives* of others—were worth less than the piles of feces and blood that those entire countries were nearly reduced to. In general, once people start stabbing or shooting one another, you won't find a lot of worry about golden rules. When the Palestinian suicide bomber is thinking about Israeli civilians, he's not thinking about it. When the Israeli settler is thinking about bulldozing the olive trees around hunger-stricken Palestinian villages, he is usually weighing neither Kant nor *Ethics of Our Fathers* nor *The Analects*.

Humanism is not, nor am I, offering anything entirely *new* here.

But while the golden rule may be simple, it is hard to follow. Religious and secular people alike fail at it all the time, and then we wonder why our lives and our countries are such a mess. And one of the reasons religion still has such a seemingly irresistible pull, to this scientifically and rationally advanced day, is that it is one of the only forces in the world whose leaders can still give themselves permission, without irony or embarrassment, to go about spouting such an obnoxiously simple—but achingly and embarrassingly important—message.

We idolize rock singers and rappers for their detachment and defiance, but rarely do they sit down with us and take the time to explain why we shouldn't get so annoyed with our mothers when they do that thing they always do to make us feel guilty. We learn Big Ideas from philosophers and other public intellectuals, but how often do they help us find the strength to be more loving husbands and wives? Psychologists and therapists will talk to us about all our problems, but they don't give warm, supportive hugs; they don't make judgments even when we want them to; and they don't come out with us into our communities and offer us positive, healthy ways to get involved with others. Clergy are among the precious few individuals in our society whose job description it is to do these things that our other heroes and guiding figures won't or can't do. A good priest, minister, or rabbi—and we've all known one or two no matter how much we might resent the religious institution that sent them our way—takes it as a professional responsibility to find ways to poke, prod, and nudge us against our will toward the golden path.

Yes, "do unto others" (or better yet, its counterpart, spoken by Rabbi Hillel—the more modestly phrased and more realistic "That which is hateful to you, *don't* do unto others; the rest is commentary") is a concept that essentially no religion misses entirely. *But not a single one of these versions of the golden rule requires a God.*

We can imagine that God forgives us for our lousy behavior, after all. Religious conservatives and liberals alike ask forgiveness of sin all the time, from Bill Clinton's famous line "I don't think there is a fancy way to say I have sinned," to the televangelist Jimmy Swaggart's slightly, well, fancier, "I have sinned against you, my Lord, and I would ask that your precious blood would wash and cleanse every stain until it is in the seas of God's forgiveness." Is anyone really naive enough to believe that such preening alone merits forgiveness of acts for which men are supposed to be damned to hell for all eternity? And yet the apologies are trotted out time and again, part and parcel of Christian morality in practice, if not according to everyone's version of Christian theory. And this pattern must embolden some who take enormous risks for the thrill of a little immoral behavior: their Lord will forgive them, if they only ask nicely enough when—or if—they are eventually caught. If you want to do something naughty, you're going to do it, and all the theology in the world isn't going to stop you.

Other people generally do not forgive us unless we earn it. And given that we have so much forgiveness to earn, imagine if we as a society put more energy into earning it. Imagine if all the arguing we do over prayer in schools, all the time we spend saying "God bless America" and "one nation under God," and all our bickering over which religion has things right were instead devoted to national days of the golden rule, and into seminars and sermons focusing solely on how we can learn to better relate to our fellow human beings—with more love and more compassion. In fact, this may be what religious scholar Karen Armstrong has in mind with her project, the "Charter for Compassion," which seeks input from people of every religious and ethical tradition—including Humanism and atheism—on how to best promote the golden rule and the idea of compassion, which Armstrong sees as its heart. Armstrong is a thinker Humanists can admire and support, and I hope many of us will join her in this effort. It may have a certain quality of Hallmark-card kitsch to it, perhaps because Armstrong seems more optimistic than is justified in believing that compassion, rather than supernatural solace and justice, is really the heart of every traditional religion. And even if such a compassion-promotion project were to be put into action rationally, on a large scale, we'd probably still miss our mark fairly often. But don't you think, despite the downside, that we'd be better off?

Of course, even if we were able to convince others to spend more time focusing on this simpler understanding of goodness and less on the vagaries and complexities of religion, we would need to provide more than one rule to live by. Sometimes people really do need a little more ethical guidance than that. So even though "the rest is commentary," as Rabbi Hillel said, let's take a closer look at what a Humanist commentary on ethical rules and values would look like. After all, Rabbi Hillel's statement comes in the Mishnah, an early, foundational book of Talmudic literature, and the Talmud—a text that could take up the entire space of some small libraries—is a whole lot of commentary.

Beyond the Golden Rule: Rules, Regulations, and Suggestions

One of the defining elements of almost every major religion is a set of moral rules, regulations, and suggestions. Judaism has Halacha, an evolving and ever-expanding set of laws that regulate or at least comment on nearly every aspect of life—though only a small minority of contemporary Jews actually pays much attention to these laws. In Islam there is Sharia, often compared to Halacha as a complex and comprehensive system of religious laws and guidelines regulating everyday life. In Catholicism, there is Canon Law. Evangelical and other conservative Christians often speak of the Ten Commandments, but in fact their tradition draws on a much broader-ranging, if less well-defined system called Christian Ethics, which has been used to attempt to regulate everything from mixed dancing to the life of Terry Schiavo to whether we can go into armed conflict with another country under the religious banner proclaiming "A Just War."

Religious laws and guiding principles are not the sole property of the West, either—one of Buddhism's founding elements was the Eightfold Path; Confucianism began with a tightly circumscribed set of hierarchies all members of a family and a state were expected to observe in relation to one another. And those were just two of the options open to Asian religionists over two thousand years ago.

All these laws were instituted so we wouldn't have moral chaos as we attempted flailingly to interpret the golden rule, or at least stop ourselves from ignoring or flouting it entirely. They were created before we had any kind of well-developed secular law—Roman democracy was occasionally nice if you happened to be a relative of Caesar, but if it got you thrown into the gladiator pit, not so much. It solved real, pressing human problems in days when there were no methods for selecting rational juries, reviewing reasonable legal precedents, or the like. And we know that even today, our legal and justice systems are painfully flawed. Any secularist or atheist who tells you that the simple solution to these problems is some glorified version of getting the word *God* off our coins or out of the pledge of allegiance is probably deluded.

But fortunately, the vast majority of secularists, Humanists, and the nonreligious would in no way claim that they have simple answers to any and all legal and moral questions. We reject the idea that any supposedly divine commandments, as they are proclaimed by human beings, ought to have absolute authority over our lives. And we believe that laws and ethical principles must come from human reason and compassion. So religious laws get a vote, but not a veto. If a given religious precept can help lead to a good life and a good society, we may adopt it. But we feel no special allegiance to laws created in an earlier time, to deal with earlier problems, according to a now-outdated value system: it is no longer necessary to refrain from mixing wool and cotton, or milk with meat, and it is unacceptable not to allow men and women, or straight people and gay people, to mix as equals in the workplace or in the place of worship, or in marriage.

How does this work out in practice? Let's begin to answer the question by looking at the Ten Commandments. In the table below I'll start with the King James Bible's version, then translate that English into modern English. And finally, we'll take a look at the Humanist version of each commandment.

THE TEN COMMANDMENTS: HOW HUMANISM COMPARES[3]

	KING JAMES	SAY AGAIN?	HUMANIST
1	I am the LORD thy God, which have brought thee out of the land of Egypt, out of the house of bondage. Thou shalt have no other gods before me.	Do not worship any other god.	Seek the best in yourself and in others, and believe in your own ability to make a positive difference in the world.
2	Thou shalt not make unto thee any graven image, or any likeness of any thing that is in heaven above, or that is in the earth beneath, or that is in the water under the earth: Thou shalt not bow down thyself to them, nor serve them: for I the LORD thy God am a jealous God, visiting the iniquity of the fathers upon the children unto the third and fourth generation of them that hate me; And showing mercy unto thousands of them that love me, and keep my commandments.	Do not make or worship idols or images of God.	Pursue truth and honesty in all you do; and be wary of allowing power, status, or possessions to substitute for moral courage, dignity, and goodness.

3	Thou shalt not take the name of the LORD thy God in vain; for the LORD will not hold him guiltless that taketh his name in vain.	Do not misuse the name of God.	Be positive and constructive rather than negative and disrespectful.
4	Remember the sabbath day, to keep it holy. Six days shalt thou labor, and do all thy work: But the seventh day is the sabbath of the LORD thy God: in it thou shalt not do any work, thou, nor thy son, nor thy daughter, thy manservant, nor thy maidservant, nor thy cattle, nor thy stranger that is within thy gates: For in six days the LORD made heaven and earth, the sea, and all that in them is, and rested the seventh day: wherefore the LORD blessed the sabbath day, and hallowed it.	Keep the sabbath day holy. You must rest on that day each week.	To be healthy, you must balance work, play, and rest.
5	Honor thy father and thy mother: that thy days may be long upon the land which the LORD thy God giveth thee.	Honor your mother and your father.	All members of the family should respect each other.
6	Thou shalt not kill.	Do not commit murder.	Same.
7	Thou shalt not commit adultery.	Do not be unfaithful to your husband or wife.	Same.
8	Thou shalt not steal.	Do not steal.	Same.
9	Thou shalt not bear false witness against thy neighbor.	Do not lie or speak badly about others.	Same.
10	Thou shalt not covet thy neighbor's house, thou shalt not covet thy neighbor's wife, nor his manservant, nor his maidservant, nor his ox, nor his ass, nor any thing that is thy neighbor's.	Do not be jealous of other people, and do not desire other people's spouses, houses, or anything else they have.	When you see nice things owned by others, let them be your inspiration, rather than a source of bad feelings. If there are things that you want, work hard to get them.

First and Second Commandments: Common Decency Is Not So Common; Ethical Excellence

One of my favorite clever little lines that Sherwin Wine delivered is this: sometimes, the nicest thing you can say about God is that he doesn't exist. Sherwin didn't mean this in a mean-spirited way. It is just that he could not in good conscience accept any of the apologetic and theological arguments (and he had heard them all) about why bad things happen to good people. Sherwin had seen the Holocaust hit very close to home. As a lifelong student of history he paid close attention to all the other man-made and natural disasters that have cruelly killed countless millions of innocent children. Nothing anyone could say about God would make him worth worshipping if he created a world that included all that senseless death. So my rabbi really did believe he was being as kind as he could to God by calling him a fictional character. As do I.

Likewise the nicest thing you can say about the Biblical first and second commandments is that they aren't meant to be taken seriously. I mean, if you were the God of all creation—of every last corner of the infinite and infinitely expanding universe; if you had created not just every nation but every person who ever lived, every species that ever crawled or swam or slithered across the surface of this planet or any other planet—would the very first ethical words out of your mouth be, essentially: "Remember that story I told you a few thousand years ago, about how I am the only God, and how a certain group of you are more special than all my other creations—you know, the story involving a massive journey out of the desert (even though there is no good archaeological evidence that it even happened)? Well, that's the reason that every human who ever lives should be very careful not to piss me off by worshipping one of my competitors, or by calling me by one of their names."[4]

I value the friendship and respect of religious people, and I didn't and don't intend this book to needlessly offend or aggrieve those friends. But the first two commandments are something that most liberal religious people just politely ignore while they go about the business of living the good lives that most of them live. Because for those of us who are primarily interested in things like raising healthy children and building a peaceful and just society, it simply makes no sense that the first words in this purportedly most important of all ethical statements, the one that people fight to have posted at courthouses and on public lawns, would not be about, well, peace or justice, or love, or compassion, or neighborliness, or anything like that. No, they are about "Worship Me. Properly. Or Else."

Of course, there are other ways to read these lines. It would be utterly irrational to think that in two thousand years of interpreting the Bible, no well-meaning person had ever come up with a rational way to understand the first two commandments that was more generous than my reading above. In fact, I'm happy to recommend some such readings, if only for cultural literacy's sake: if you want to read impressively intelligent Biblical interpretation by modern religious intellectuals, see folks like Jim Wallis, N. T. Wright, Karen Armstrong, Tony Campolo, Peter Gomes, Serene Jones, Michael Lerner, Jonathan Sacks, Judith Plaskow, and many others. I've read their works and they contain much impressive wisdom.

But nobody can convince me that these are the messages that belong at the beginning of a series of the ten most essential commandments I ought to teach my children about how to be a good person.

So what would Humanists teach their children to open such a fateful lesson? In fact, the chart above is adapted from a Humanist Sunday school lesson, so let's take a closer look at its answer:

Seek the best in yourself and in others, and believe in your own ability to make a positive difference in the world.

Pursue truth and honesty in all you do; and be wary of allowing power, status, or possessions to substitute for moral courage, dignity, and goodness.

The first thing Humanists would teach their children is the same thing many religious people would probably say they would teach: seek the highest, most worthwhile things you can find. We just disagree in good faith about what those highest things are. We do not seek ourselves—this is key—but rather, we seek the best in ourselves. We seek dignity for ourselves, and the best way to do so is to seek to help others achieve dignity. If you seek only yourself, at best you will find a mixed bag. Some days you look into the mirror and see a confident person, some days an insecure one. Some days you feel kind and loving. Other days, if you asked yourself, "What did I feel today that was kind and loving?" you might very well come up with nothing. But every day, like the Buddha on his journey, we come across opportunities to offer our best—or our worst. We are surrounded by people who are sick, aging, dying, or grieving for loved ones who are sick, aging, dying, or dead. These are the hallmarks of an unfair world, and Siddhartha was traumatized by them because he, like any of us, wanted to heal the pain he saw in himself and others but realized he could not do so. So he turned inward to eliminate his own *wanting* (which of course many did see as help, as they too saw the appeal of ending the endless cycle of appalling and repulsive things).

The Humanist response is different. When people are killed, we seek the best ways to remember their lives and legacies. If they were mainly good, beloved people, we try our best to carry on the good things they did. If they were ineffectual people who more often than not did not know how to love, we try our best to learn from their mistakes, rather than simply hating and resenting them now that they are gone. If they were too young to be either bad or good, we acknowledge our special pain and anger at the loss of the hope that their new lives brought. We do not try to paper over the deep wounds by saying, "They are with God now." We try to comfort one another, to offer hugs, kisses, time, patience, and presence, because no supernatural force can offer these things, and we need them. We remind ourselves, again and again if necessary—and it usually is necessary—that we are capable of making a positive difference, even if a small one. We remind ourselves that in the face of philosophical talk about string theory and the size of the universe it can feel as if we make no difference, but that each time we offer just one gentle caress of the face of a loved one in pain we make a great difference indeed.

Consider this poem by Yehuda Amichai:

Roshi, Roshi—when I banged my head on the door
When I banged my head on the door, I screamed,
"My head, my head," and I screamed, "Door, door,"
and I didn't scream "Mama" and I didn't scream "God."
And I didn't prophesy a world at the End of Days

where there will be no more heads and doors.
When you stroked my head, I whispered,
"My head, my head," and I whispered, "Your hand, your hand,"
and I didn't whisper "Mama" or "God."
And I didn't have miraculous visions
of hands stroking heads in the heavens
as they split wide open.
Whatever I scream or say or whisper is only
to console myself: My head, my head.
Door, door. Your hand, your hand.[5]

This is one of those poems that bring me as a Humanist about as close as I can come to the feeling an earnestly devout Christian might get reciting the Lord's Prayer. I consider it liturgy, and I could just as easily have included it in chapter 6 on Humanistic alternatives to traditional prayer. For now, however, a key insight of Humanism is that we do not have to make more or less of our human experiences than what they are. When we feel pain, it is neither an illusion nor a sign from the heavens. It is not God, and it is not the Devil. It is pain, sadness, and longing. Running away from it, pretending it does not exist, will do us no good. But often an almost magical thing happens when we stop to acknowledge our pain: we are more easily able to let go of it and focus on simple pleasures.

In recent years, Positive Psychology, the direct contemporary descendant of the Humanistic Psychology movement of the mid-twentieth century, has designed a concrete tool for helping us to seek the very best in ourselves. Psychologists Martin Seligman and Chris Peterson have pioneered the idea of an "inventory of strengths," a well-organized, comprehensive list of the various types of positive character traits human beings can possess. Among the benefits of this inventory is that it is an alternative to a view of psychology that focuses only on all the little things that are wrong with us. For decades, the practice of clinical psychology has been dominated by a doorstop of a book called the DSM (*The Diagnostic and Statistical Manual of Mental Disorders*), an epic catalog of all the things, big and small, that can go wrong with our minds. Of course, it can be important to understand and treat real psychological problems, but at times we can become so focused on our various inadequacies that it is no wonder we feel we need an all-powerful God to rescue us from them all. The truth about us is often more balanced.

Seligman and Peterson looked at every description of human virtues they could get their hands on, "from the holy books of major religions down to the Boy Scout Oath ('trustworthy, loyal, helpful, friendly')."[6] (You can read more about the process and their reasons for undertaking it in the chapter titled "The Felicity of Virtue" in Jon Haidt's *The Happiness Hypothesis.)* Seligman and Peterson came up with a list of six broad virtues, or clusters of virtues, that came up again and again in many types of lists from many different sources. They identified the six families of virtues they found this way: wisdom, courage, humanity, justice, temperance, and transcendence.

One can argue all day about which other positive qualities should have made the list, or how it should have been divided up differently. The point, as Haidt also suggests, is that we Humanists should try to avoid beating up ourselves or the people around us over which qualities we lack. We don't complain or harp if someone with the courage and determination of a Lance Armstrong doesn't also turn out to have the laid-back peacefulness of a . . . well, actually, I had trouble identifying an example of a famous laid-back, peaceful person, because extremely laid-back people very rarely have the drive and intensity needed to achieve fame—and that's okay too!

None of us will ever be all things to all people, or to ourselves, but we can identify which areas we're particularly strong in and cultivate those talents, surrounding ourselves with friends, loved ones, and coworkers whose strengths complement ours.

It is worth mentioning, however, given that Peterson and Seligman encourage us to debate and discuss the qualities on their list, that the idea of honesty, or truthfulness, does not appear on it. For Humanists, honesty is a core, cardinal virtue. This is true both in the sense of having the honesty to reveal the truth to others even when it is difficult (we don't obsess about little white lies, but finding a better way to be truthful is preferable—for example, "Thanks, I'll think about what you're saying" is better than a blatantly false "Thanks, that's a great idea!") and also in the sense of facing reality, being honest with ourselves about what the world is really like, while having *faith* that we will be able to handle the truth. The fact that this is so important and so difficult is one of the main reasons I'm so comfortable calling Humanism a faith.

In our upholding of honesty and truthfulness as a virtue, we Humanists admire the Biblical character of Abraham, who smashed his father's idols to false gods. We see scientists as idol smashers when they help us overcome the oppression of yesterday's knowledge and assumptions. Psychotherapists are idol smashers when they help us overcome irrational fears about ourselves, assumptions that we are no good, or that we always have to be the best at everything. And any of us can be idol smashers when we refuse to give in to the idolatry of crass commercialism and its message that financial wealth and conspicuous consumption always equal happiness—though we'll explore that more later.

Abraham's story is a great Biblical theme that we can draw inspiration from to this day, though we go at least one idol further than Abraham would have, smashing as well the idea that even one God is necessary in order to live well or find meaning in the world around us. And we also recognize that there are plenty of non-Biblical sources of similar inspiration, and that the Bible's story is more than a little self-contradictory. After all, this same Abraham was also willing, without even a question, to slaughter his own beloved son in order to propitiate God (not a virtue likely to appear on *any* version of Seligman and Peterson's list).

Third Commandment:
Winners Don't Punish, or Nice Guys Do Finish First

Thou shalt not take the name of the LORD thy God in vain: for the LORD will not hold him guiltless that taketh his name in vain.

vs.

Be positive and constructive rather than negative and disrespectful.

The first part of the Biblical third commandment is unfortunately not even worth dealing with seriously at this point in history, though at a time when people really did believe that competing invisible gods could become angry on a moment's notice and send angry natural disasters or other misfortune our way, I suppose it was important enough to remind everyone within spitting distance not to risk pronouncing a curse. Today, however, one shudders to think what would become of the sex talk in our "very religious country" if people suddenly began to take seriously the commandment against taking the Lord's name in vain: Christian America would need a whole new bedroom vocabulary. When I wrote an essay supporting a removal of mandatory mealtime prayers at the Naval Academy for the *Washington Post/Newsweek* blog On Faith, one truly pious soldier wrote in to say that he would gladly accept a ban on mentioning Jesus in public prayers before meals, if only we could also ban the much more frequent invocation of his name *during* meals as part of all manner of crude expressions.

In the second half of the third commandment the Lord promises to punish all who commit the above-discussed minor infraction. Set aside the obvious humor in the thought of how deserted heaven would be if God actually punished all who take his name in vain—I'll leave that one to Louis Black or the late, great atheist comedian George Carlin, god rest his soul. Simply notice that God's strategy is: when in doubt, punish the wicked. Take a moment and ask yourself, "When have I adopted this strategy in my own life?" How did that work out for you?

Some of my colleagues at Harvard recently performed an ingenious study that involved observing the tendency to punish among several hundred people playing the classic "prisoner's dilemma" game often used in social science research. The rules of the simulation game are simple: two "prisoners" can either choose to help each other in attempting to "escape," or they can adopt a more winner-take-all strategy. Collaboration is the more effective means of escaping but requires trust that your coconspirator will not betray your trust in an attempt to escape without you. The researchers looked at the behavior of players who were repeatedly betrayed: did they choose to retaliate, when given the chance, against those who'd double-crossed them?

The study's findings were heartening and decidedly non-Biblical in their message. Players lost in the long run if they responded to betrayal by going out of their way to punish their opponents. In the long run, as the researchers titled their study, "Winners Don't Punish."[7] Nice guys do finish first, because winners are not bitter, and bitter people are not winners. *This* is a message truly worth teaching to our children, and to each other. When you are slapped, you don't necessarily have to turn the other cheek to be slapped again, but neither should you slap right back. You can walk away. You can work hard to move forward without wasting your own precious energy making sure that those who cause you suffering suffer in turn. Let the impulse go.

Recall, for example, that the Marshall Plan worked out a lot better than the Treaty of Versailles. After we concluded an unspeakably brutal World War II with Germany and Japan, what did we do? We offered our vanquished enemies all manner of money and assistance. We supported them! And the results have been positive, relatively speaking. Contrast this with the disastrous Treaty of Versailles: After we beat Germany in World War I, we imposed conditions that amounted to lording it over our enemies and rubbing their collective noses in their vanquished aggression. Which of course led to a fight against them all over again, and they had become angrier and more aggressive the second time. This principle can be extrapolated—to Israel and Palestine, which have been punishing each other continually for half a century of traded wrongs to no avail—or to Cuba, which we've been punishing for the sin of Communism for fifty years. The idea seems to be if we punish them enough, they'll come to their senses, but they haven't yet. Half a century.

This also goes for how we should deal with the Middle East as a whole. The Israeli multibillionaire business magnate and philanthropist Steff Wertheimer has worked for years to try to convince his own country, as well as the United States, the Palestinians, and anyone who will listen, that we need a "Marshall Plan for the Middle East." Wertheimer's idea is simple but beautiful. Few if any Middle Eastern countries can claim a sustainable, above-poverty standard of living for average people. And while it is true that not all terrorists come from impoverished backgrounds, it is undeniable that if there were not so many people suffering destitution in the Middle East, there would be less fuel for anti-Western hatred. If we could help these nations create sustainable, peaceful industries—perhaps even sources of green energy so they could become independent from us even as we seek independence from them, rather than simply funneling oil revenues that can easily be co-opted by corrupt governments—we would do more to turn back the tide of terrorism than any number of "smart bombs" could ever do.

Positivity and constructiveness are not just for politics, however. This approach is incredibly helpful in our personal lives as well, and most of us, religious or not, are in almost constant need of a powerful reminder to take a "Marshall Plan" approach to our loved ones as well—which is why a certain *New York Times* "Modern Love" essay, "What Shamu Taught Me About a Happy Marriage," sat at the top of the *Times*'s "most e-mailed" list for day after day in the summer of 2006.

In "Shamu," writer Amy Sutherland discusses candidly how her loving marriage to her husband, Scott, was slowly breaking down, not from infidelity or gross incompatibility, but by the death of a thousand cuts, a thousand little annoyances. That is, until an assignment to write about animal trainers woke her up to a different outlook:

> The central lesson I learned from exotic animal trainers is that I should reward behavior I like and ignore behavior I don't. After all, you don't get a sea lion to balance a ball on the end of its nose by nagging. The same goes for the American husband.
>
> Back in Maine, I began thanking Scott if he threw one dirty shirt in the hamper. If he threw in two, I'd kiss him. Meanwhile, I would step over any soiled clothes on the floor without one sharp word, though I did sometimes kick them under the bed. But as he basked in my appreciation, the piles became smaller.[8]

Obviously you can take this approach whether you believe in God or not. In fact, I have no idea whether Sutherland considers herself a Humanist, and I don't care. Because the point of Humanism is not just to collect little factoids, nuggets of knowledge about what we're supposed to do and who tells us to do it. It takes more than reading some self-help article—or some Biblical passage for that matter—to become wise. The point is, we need to take concrete steps to put these ideas into practice in our daily lives, and to solicit support from those around us in doing so. So in our Humanist community at Harvard, we try to spend a lot less time staging debates between Christian fundamentalists and atheist philosophers, or even lectures and symposia on the evils of "intelligent design," and more time on adult education seminars where we study subjects like Nonviolent Communication, the technique Marshall Rosenberg has articulated for learning how to speak to others—and to ourselves—more directly and compassionately. Rather than complaining about how much power religion has in today's society, we want to study how to empower ourselves to respond to all the problems in life—aging, pain, death, and the myriad stresses in between—that make people turn to religion in the first place. Humanists don't believe in turning water into wine—but we definitely believe in turning lemons into lemonade.

Fourth Commandment: Lifestyle Balance

Remember the sabbath day, to keep it holy. Six days shalt thou labor, and do all thy work: But the seventh day is the sabbath of the LORD thy God: in it thou shalt not do any work, thou, nor thy son, nor thy daughter, thy manservant, nor thy maidservant, nor thy cattle, nor thy stranger that is within thy gates: For in six days the LORD made heaven and earth, the sea, and all that in them is, and rested the seventh day: wherefore the LORD blessed the sabbath day, and hallowed it.

vs.

To be healthy, you must balance work, play, and rest.

We religious Americans don't seem to be following any version of the sabbath commandment, do we? The average American is lucky to have more than a couple measly weeks of paid vacation per year. Ask secular Parisians what they think of that—but don't ask during the month of August, because almost all of them will be lying around on the Riviera somewhere enjoying a few of the seven or eight weeks of annual paid vacation the French take "religiously." We Americans weren't always such workaholics. As recently as the 1960s, Europeans worked longer hours and took less vacation than we did, but divergences in labor union negotiations, tax policy, and culture (depending on which economic theory you accept) caused Europe to become, apparently, the better keeper of the spirit—if not the letter—of the traditional fourth commandment.[9] Which leaves one to wonder: could we have become a more religious country in recent decades in part because we want God to save us from our self-imposed overwork?

And speaking of workaholics: any kind of addict who is uncomfortable with or not interested in focusing on the "higher power" found in twelve-step programs can turn to a wonderful secular alternative, designed by

Humanists, called SMART Recovery (SMART stands for Self-Management and Recovery Training). The physicians and psychologists who designed SMART, in accordance with the best, most empirically demonstrable science about addiction and recovery, all agree that one of the key elements for success in addiction recovery is what they call "lifestyle balance."

Lifestyle balance, in the context of recovery, means that we don't become addicted to drugs, alcohol, gambling, or other negative behaviors unless they actually have some benefit for us in the short term. So, people may start drinking heavily because it lessens their inhibitions and allows them to ignore feelings of discomfort in social situations. Eventually, though, many alcohol addicts come to talk about the bottle as their mistress, best friend, and only true companion. Drinking becomes the one and only time when they can feel pleasure, or at least escape some very painful feelings. Given this context, learning to successfully abstain from drinking has to be about more than just not picking up a drink. Long-term recovery requires finding healthier and more sustainable ways to feel good, relieve stress, and enjoy life. SMART Recovery patients and clients, I was initially surprised to discover, are encouraged not just to count their days of sobriety and stay away from bars and pubs, but to explore new hobbies, sports, and friendships and to balance work with relaxation and family time.

As Dr. Joe Gerstein, emeritus professor at Harvard Medical School and one of SMART's founders, put it: "This is the fundamental equation in the recovery from any sort of addiction or preoccupation: balancing immediate and enduring satisfactions. How much do you sacrifice in relationships in order to succeed in career goals, etc.?"[10] Gerstein added in an interview,

> During addictions of any kind, life clearly falls out of balance, not only short-term vs. long-term concerns, but in terms of daily time expenditure. A cocaine or heroin habit requires a tremendous involvement in planning and executing a buy, thinking about being and getting high and financing purchases and, of course, actually being high. When this activity ceases, there is a huge hole to fill with activities, distractions, relationships (and possibly repair of relationships), etc. So there is this time rebalancing that has to go on. A man in one of my groups who has a gambling addiction can't

> rebalance right now because he is working 14–16 hours a day, 7 days a week in order to repay his lenders of various sorts, but [he] is well aware of the need to attend to other areas of life once his finances get under control. Likewise, we have many people involved with SR who were energetic, competent professionals and business people who retired and thought things would just take care of themselves but instead began drinking more and more because they didn't fill up the hole left by retirement with other sorts of endeavors and distractions.[11]

I've enjoyed visiting SMART meetings very much. Though I've personally never had substance abuse issues, and initially only wanted to go as part of researching this book, they've given me a great deal of insight into my occasional bouts of workaholism and Internet addiction. We spend so much time glued to the computer screen these days, I wonder if half my students couldn't use a support group that would help them get out and interact with one another in real time rather than online. If they're not addicted, why do their palms get sweaty when they have to go without e-mail for a whole day?

The point that I find sad and maybe a little bit absurd is that, on the one hand, we've reached the point where our commitment to Judeo-Christian religious principles is so superficial that the overwhelming majority of Christians and Jews in the United States pay zero attention to the Biblical commandment to rest at least one-seventh of the time. And yet on the other hand, because we are so interested in giving lip service to our religiosity and so afraid to openly acknowledge our newer, more Humanistic value system, we almost never take time to acknowledge that our Biblical ancestors may have had an important insight about the need to balance rest and work! We need to figure out how to rest in this fast-paced, high-tech society, but just telling everyone to sleep in on Sunday is not a viable solution. It takes, well, work to get this right, and we'd be better off if we could just talk directly with one another about the challenge and about how to solve it together. Some people recoil from a Humanist outlook on morality, both around this issue and around others, because, as we discussed in chapter 1, they see Humanism as self-interested. But as Gerstein likes to say,

Humanism emphasizes that you should look out for your own interest . . . your own *long-term* interest. This is the key.

Fifth Commandment: Humanist Family Values, Gay Marriage, Etc. Etc.

> *Honor thy father and thy mother: that thy days may be long upon the land which the LORD thy God giveth thee.*
>
> vs.
>
> *All members of the family should respect each other.*

What the American right wing calls "traditional family values" actually make some sense—within the context of an ancient agricultural society. We often forget that until approximately ten thousand years ago, humans had not yet invented agriculture. For most of the time that humans have walked the earth, we got our calories exclusively from hunting and gathering, so life was a day-to-day experience. Nearly everyone needed to be devoted to the enterprise of basic survival. Slowly, this began to change when we learned techniques for producing our own food. Agriculture meant we could predict when and where we would be able to survive and thrive, months and years in advance. But farming, along with increased education to pass on ever-increasing knowledge about agricultural techniques, also required more manual labor. To ensure that there would be enough men to work the fields, fertility had to be valued as never before. Agricultural families urgently needed to produce reliable helpers. This made heterosexuality a valuable commodity. Wasting semen was to be frowned upon. Killing babies or inducing the abortion of a fetus would have been the greatest sin. Family members had to know their place, because agricultural life thrives on stability and predictability.

Such a moral system, as it emerged from the Fertile Crescent and other early cradles of crop production, was never universally popular and sometimes caused serious conflict. As mentioned earlier, the Chinese Confucian tradition—an example of the farmer morality—initially offered fierce resistance to the coming of Indian Buddhism with its antiagricultural insistence on monkish celibacy and asceticism. But the agricultural ethic was well suited to huge swaths of relatively arable territory in Europe and North America, and it stuck with us for hundreds of years here in the West, enmeshed with Christianity, as the dominant ethos—until the secular revolution began to threaten its dominance. In the increasingly modern, secular cities that have emerged in modern times and places where only Christian farmlands once lay, morality has changed as much by necessity as by design. An unlimited number of children is no great advantage when living in cramped, overcrowded communities. Knowing your place is no longer as good an idea when new knowledge and technology are in constant demand, and whoever can provide them will be valued, regardless of his or her station in life.

Humanism embraces the morality of the secularized, urbanized, interconnected world, even with all its uncertainty. We accept with enthusiasm the modern proposition that all people must be free to make basic choices about the shape of their family life: whom to love, whether to have children, how to structure a family. And we recognize that neither strength nor nurturing is in enough supply to be confined to one gender or one type of person. Humanists are *progressives*, meaning that, though we may disagree in good faith on particular economic, security, or social policies, we believe we have the opportunity and the responsibility to help make *progress* toward a world that will be fairer and more just, more loving and accepting of difference than the world we were handed by the fates.

As a progressive, one of the first things I want a more powerful and prominent Humanist community to address is those very same "family values" that the religious right has stolen from us. Their distorted wish is to return to their version of an old morality of festering homophobia, obsession with controlling women's bodies, and harsh, often hypocritical discipline. Let's steal family values back by actively celebrating our understanding of the diversity of what a family can be and our more welcoming, nurturing approach to life!

Valgard Haugland, Norway's Christian Democratic minister for children and family, has said: "Americans like to talk about family values. We have decided to do more than talk; we use our tax revenues to pay for family values."[12] Humanists may have disagreements about how to use tax money, but we too like to do a lot more than *talk* about building healthy families. We get down into the trenches and work at it.

Love Me Do: Loving Behavior

On an individual level, Humanist family life is built on the foundation of loving behavior. The word *behavior* here is the key. Both Humanism's supporters and its critics tend to understand that we are against rigidly predetermined roles for men and women. But Humanists and anti-Humanists alike also tend to assume that the alternative we propose is a life based on *love—romantic love,* the kind the Beatles sang "Love Me Do" about, the kind Hollywood seems to make one bad movie after another about. They don't see that such love is only part of the equation, and this is not only why some of our conservative enemies hate us so much, it is probably why so many of our marriages struggle to the point of breaking. Because to base our entire family life, and thus our entire society, which after all is made up of family units, on an emotion that changes, appears, and disappears so often that sometimes it seems we're able to employ an entire Greenwich Village full of artists, novelists, and musicians trying to pin love down and describe what the hell it is and where it goes when it's gone, is, to say the least, risky.

I don't care how "rational" you consider yourself when it comes to religion, if you expect to find a feeling of passionate love for another person, regardless of gender, and based solely on that *feeling* conceive children and raise them from needy infancy to well-adjusted adulthood while maintaining an active, maybe even lightly and playfully kinky sex life, a lovely home with a well-manicured lawn, a balanced checkbook, a successful career, and great shoes, you're most likely running on the steam of a delusion far stronger than the stuff they bake into run-of-the-mill belief-in-God brownies.

Feelings change, and their changes are not under our control. It isn't possible to feel the same way about one person for your entire life. Novelty, variety, and mystery will always be tempting. That is the scary news. The good news, however, is that our need for companionship, touch, compassion, and trust does not change. These are things we can best enjoy by maintaining long-term monogamous partnerships. We choose partners because we feel a sense of excitement around them, a thrill to be near them, but we also recognize that these feelings are not magical and are certainly possible to feel for multiple people. So we choose wisely: not someone who can meet all of our needs, because no such person exists, but one who can complement us for the long term as well as the short term—and then we choose to continue to nurture that partner both when the feelings of passion are there and when they are harder to find. Passion is not magic or God-given—it may seem to vanish, but it can return if we allow it to. Maybe it's not the stuff of some junk-food-sweet fairy tale, but it's the truth. And that's what I call "true love."

Gay, Lesbian, Bisexual, Transgender, and Other Nontraditional Families

It should almost go without saying—but let me say it again anyway—that Humanism is in full support of gay marriage and parenting. I thought about adding a special section here for gay Humanist parents, but it is not something on which I'm an expert, and besides, the central message is that Humanists consider such families 100 percent equal to heterosexual ones—no better, no worse. Just human.

I'm aware that we're not alone in doing so. In North America, for example, liberal churches and synagogues have opened their doors and their theological interpretation wide to the gay community, and that's a good thing. I also understand why a lot of gay and lesbian couples I know are anxious to have their partnerships recognized by the religion of their parents and grandparents: in a society where something as fundamental as whom you're attracted to can mean that you're still considered at least a little bit of an outsider, it can be comforting to feel you are gaining the approval and blessing of God and his prophets and priests.

There are those who wonder how it is that any gay people are religious at all. They expect that, because of the way conservative religion has treated homosexuality over the last many centuries in the West, and because liberal religion adopts many of the rituals and trappings of conservative religion, even if not the whole of the dogma, the vast majority of GLBT people would be self-identified atheists or agnostics. This strikes me as fanciful, however. We talked earlier about the evolutionary reasons for the belief in God. Well, these affect gay people just as much as straight people, even if there are some small differences between the brains of gay and straight people, and as far as I know, there's no evidence yet that such differences exist.

Still, I wonder if you'd forgive me a moment of pride in the fact that Humanism is one tradition in which it is neither a minority view nor a recent development but an almost universally accepted fact of basic outlook and policy to afford equal respect and consideration to gay spouses and parents, as well as single parents, intercultural, interracial, and interfaith couples, and anyone else out there trying their human best to live a decent, loving family life. Family life isn't easy: In fact, I've always wondered why social conservatives don't adopt the attitude expressed in the joke about a right-wing Christian who announces to his friend that he's pro–gay marriage. When the friend incredulously asks why, the first says, "Why should we be the only ones to suffer?" But seriously, our goal is to support one another in enjoying the best parts of family life by strengthening both family and community ties—whatever and wherever they might be.

And as a final point for this section, when we say all members of the family should respect each other, it does not mean overlooking the differences between parent and child. It does not mean that children should be given everything they want or that they should be able to do whatever they want to do. It is rather a recognition that parents are fallible human beings who should earn the respect of their children by providing consistent caring, by meeting those needs that they are capable of meeting, and by showing over time that they use the word *no* selectively as a tool for furthering their children's long-term interests. Parents and children who adopt this model often find that they become close, even best friends, over the years. But for much more on the subject of *Parenting Beyond Belief*, please see the wonderful book and Web site of that name, by award-winning author, academic, and lecturer Dale McGowan, who has become known as a national expert on the subject of parenting for Humanists, atheists, agnostics, and the nonreligious.

Sixth, Seventh, Eighth, and Ninth Commandments: Rights from Wrongs

Thou shalt not kill.
Thou shalt not commit adultery.
Thou shalt not steal.
Thou shalt not bear false witness against thy neighbor.

As leading twentieth-century Humanist philosopher Corliss Lamont once wrote, "We cannot stress too much the cardinal importance of plain, old-fashioned honesty in every walk of life."[13] Humanism affirms these four of the Ten Commandments without abridgment. I wonder if that says something about the relationship between Humanists, or the nonreligious, and traditional religion in general—that we agree with just under half of them. I don't know that any such generalizations should be made here, and even if so, I'd be unsure of their meaning. After all, as a baseball lover, I know that having a successful outcome in four out of ten attempts is in some fields considered not merely good but all-time great.

Still, while it's not important or (I hope) necessary to spend any further time explaining that it is every bit as plain for nonreligious people as for religious ones to recognize that murder, adultery, thievery, and lying—among many other such pernicious activities—are wrong, we might profit from some discussion of *how* Humanists go about determining which such actions are worthy of censure and which are punishable offenses. We've talked about the obnoxious question of whether we have a method for determining right and wrong generally. The good is that which facilitates human dignity and the health of the natural world that surrounds us and sustains us. The bad, or evil, is that which creates needless human suffering. This, then, leaves us with the sometimes difficult but always worthwhile task of debating what should be considered human dignity and health, and what should be considered needless suffering. Or, in the words of Erich Fromm, "Humanistic ethics, for which 'good' is synonymous with good for man and 'bad' with bad for man, proposes that in order to know *what* is good for man we have to know his nature. *Humanistic ethics is the applied science of the 'art of living' based on the theoretical 'science of man.'* "[14] Contrary to the notions of those who would demonize Humanists and the nonreligious out of ignorance, fear, or outright hatred, we do not have a very hard time determining that murder, adultery, theft, and lying are bad, not only because we can see the effects they have on people, but most likely also because we evolved to have a sense, regardless of which religious beliefs we do or do not hold, that such things are wrong.

Today, a new field has emerged that combines psychology, philosophy, and hard science to study the evolution of morality. This research is producing fascinating information; it would take more space than we have here to summarize the work being done by cognitive scientists such as Steven Pinker, Marc Hauser, Joshua Greene, and dozens of other researchers to determine just how human beings came to have beliefs and intuitions about morality and ethics. Two points are crucial to keep in mind for now, however.

First, our reasoning about basic moral questions is often a lot more complicated than it seems, but it does follow basic patterns. This is best understood by observing an experiment designed by philosophers Philippa Foot and Judith Jarvis Thomson, called the "trolley problem." Imagine a trolley car barreling down its track, out of control, toward five unsuspecting workers. The conductor has passed out. The workers will not see the trolley in time. And you are at a switch with the opportunity to divert the trolley's path. Only one problem: the trolley would then kill one lone worker standing on the other track. Would you pull the switch anyhow? Most people say they would. But what if, instead of a switch, you were standing on a bridge above the trolley's path and could stop the trolley by pushing a heavy object into its path? And what if the only heavy object available were a fat man standing next to you on the bridge? Most people say they would not make the second move, even though its result is exactly the same as the first—killing one person to save five.

Second, there is overwhelming data to suggest that human beings tend to respond similarly to the trolley problem and other similar situations, regardless of religious belief or affiliation or lack thereof. As Steven Pinker put it in the *New York Times Magazine*:

> When psychologists say "most people" they usually mean "most of the two dozen sophomores who filled out a questionnaire for beer money." But in this case it means most of the 200,000 people from a hundred countries who shared their intuitions on a Web-based experiment conducted by the psychologists Fiery Cushman and Liane Young and the biologist Marc Hauser. A difference between the acceptability of switch-pulling and man-heaving, and an inability to justify the choice, was found in respondents from Europe, Asia and North and South America; among men and women, blacks and whites, teenagers and octogenarians, Hindus,

Muslims, Buddhists, Christians, Jews and atheists; people with elementary-school educations and people with Ph.D.'s.[15]

What this brilliant and brilliantly executed experiment shows is something those of us who know a lot of atheists and a lot of religious people could have told you was obvious: there really isn't much difference, on a practical or everyday level, in how members of these two groups answer moral questions such as whether, and when, it is right to kill.

It will be interesting to observe the impact of this new moral science on human rights. Ever since the planning of the 1948 United Nations Universal Declaration of Human Rights—and indeed, since the dawning of modern times—scholars and political and social leaders have been debating the source of the very concept of human rights. Are such rights given to us by God, and is that the only way we can account for their existence? Or is there some secular way of understanding human rights such that they will be taken every bit as seriously as if they had been handed to us by a deity? Leave it to the firebrand law professor Alan Dershowitz to answer affirmatively to the latter question in a tone that would make Richard Dawkins and Christopher Hitchens proud: "Alas, the claim that rights were written down by the hand of the divinity is one of those founding myths to which we desperately cling, along with the giving of the Tablets to Moses on Sinai, the dictation of the Koran to Muhammad, and the discovery of the Gold Plates by Joseph Smith."[16]

In a compelling book called *Rights from Wrongs,* Dershowitz rejects Aristotle's claim that we can't know what rights we have unless we know what kinds of lives we ought to live, because, he argues, though we'll never agree on whether it's better to be a libertarian or a tax-and-spend liberal, we can agree on certain things we don't want to see happen: such as another holocaust or genocide. Similarly, even though we Humanists won't always agree, not only on whether to call ourselves "Humanists" or by some other name, but on what human dignity (or whatever word you prefer to refer to what I'm calling dignity) should look like—we can all agree that we want to reduce needless human suffering.

Dershowitz is hardly alone among human rights scholars in arguing for a secular conception of rights. Nor is he the least controversial in other areas of thought. Still, *Rights from Wrongs* is worth reading because it lays out perhaps the most systematic vision yet for just how we Humanists and secular people think about these issues. As he summarizes it: rights *do not come from God*, because God does not speak to human beings in a single voice, and rights should exist even if there is no God.

> Rights *do not come from nature*, because nature is value-neutral.
>
> Rights *do not come from logic*, because there is little consensus about the a priori premises from which rights may be deduced.
>
> Rights *do not come from the law alone*, because if they did, there would be no basis on which to judge a given legal system.
>
> Rights *come from human experience*, particularly experience with injustice. We learn from the mistakes of history that a rights-based system and certain fundamental rights—such as freedom of expression, freedom of and from religion, equal protection of the laws, due process, and participatory democracy—are essential to avoid repetition of the grievous injustices of the past . . . In a word, *rights* come from *wrongs*.[17]

Tenth Commandment

> *Thou shalt not covet thy neighbor's house, thou shalt not covet thy neighbor's wife, nor his manservant, nor his maidservant, nor his ox, nor his ass, nor any thing that is thy neighbor's.*
>
> vs.
>
> *When you see nice things owned by others, let them be your inspiration, rather than a source of bad feelings. If there are things that you want, work hard to get them.*

Oh, "thy neighbor's wife," how unhelpful a commandment you are. Let's remember that the subject here is not adultery, which we agreed to condemn in commandment seven, but coveting itself. The extent of the Bible's unhealthy attitude toward covetousness can be most fully appreciated when we read the tenth commandment alongside this statement, purportedly by Jesus, from the New Testament:

But I say unto you, That whosoever looketh on a woman to lust after her hath committed adultery with her already in his heart. And if thy right eye offend thee, pluck it out, and cast it from thee: for it is profitable for thee that one of thy members should perish, and not that thy whole body should be cast into hell. (Matthew 5:28–30)

Tolstoy has a story, in the same novella as *Ivan Ilych*, about a young nobleman named Evgeny who has a liaison with a married peasant girl, Stepanida, when he is a bachelor. Evgeny and Stepanida sleep together a number of times; he then breaks off the affair without thinking much of it. Apparently hers was a troubled marriage and he was not her only affair. A year later, Evgeny is married to a woman he thinks very highly of, but a little while into his marriage, when his young bride is pregnant, Evgeny sees Stepanida again and suddenly finds himself thinking of her constantly. Rather than recognizing that it is natural for him, as a still young man with many responsibilities, to have fantasies about a life and pleasures beyond his own, he becomes racked by guilt that he would even entertain the notion of sex with his former lover. Evgeny takes the morality of Matthew 5 as his model (the story is prefaced by verses 28–30 to underscore this point) and treats the situation as though he has already sinned, and moreover is already a terrible sinner. Evgeny never again so much as touches Stepanida, and yet because of his conviction that his fantasies are as good (or rather, as evil) as having acted on them, he allows his thoughts of marital transgression to ruin his life in the most unexpected and tragic ways.

Too many of our lives are still marked by obsessions like Evgeny's. We don't know how to forgive ourselves for our human desires, for our inability to squeeze our fleeting thoughts into the same consigned spaces as our concrete commitments. And so we are constantly disappointed in ourselves and in others around us, rather than either celebrating the vitality of our imagination or, even more importantly, getting to the work of disciplining our behavior so that it does not follow helplessly after natural but nonetheless impractical thoughts.

Coveting in itself is not a sin. Rather than endless "theological interpretation," which in cases like this is really little more than the religious equivalent of all the spin and talking points modern political campaigns use to manipulate truth rather than honor it, can't we just acknowledge reality for once? Jesus may be a compelling cultural symbol. He may even have been a real historical personality with insights on a range of issues. But Jesus was wrong on this issue. We should *not* pluck out our eye if looking at things we want seems to be causing us to "sin" to get them. We should examine our *behavior*, and work on adopting a more realistic attitude about how to attain the things we want, and whether they are really worth attaining.

New Commandments?

It can be enjoyable and enlightening to do this sort of close study of traditional religious texts and our Humanistic responses. But in looking for an ethical code for today, we shouldn't necessarily think we can get by through merely formulating responses to the Ten Commandments. Doing so would imply, after all, that the early Jews and Christians had their ethical priorities straight in a way that they did not. Certainly the vision of what was ethically best for a group of priests or rabbis two millennia ago shouldn't have the final say on what issues are most ethically pressing today. The above issues may all be very important, but they are not necessarily the definitive Top Ten most important points to remember in life.

What would a Humanist's ten principles be? Here are a few we shouldn't miss:

Humpty Dumpty and the Growing Garden

Humanism is in many ways a response to one book, one very old book that purports to have a moral message, and this purported message has in truth reached and influenced millions. And the message is wrong, and frankly pernicious, because it's causing us to live our lives wrongly, build our society wrongly, and we *should* be angry about this book. Maybe not at the author or even the readers—they know not what they do—but certainly anger at the book itself is appropriate. We need to fight against its message until that message has been defeated. And of course you all know what book I'm talking about, don't you?

Humpty Dumpty. That's right, *Humpty Dumpty*. You know the story: "Humpty Dumpty sat on a wall. Humpty Dumpty had a great fall. And all the king's horses and all the king's men couldn't put Humpty together again." The End.

There is a Humpty Dumpty mentality that says that the world—be it our personal lives or society as a whole or whatever—needs to be *repaired*. That things were once perfect and round and bright and shiny like an egg until they fell and broke into a million pieces, and now it's our job to reassemble all the pieces.

The only problem with this mentality is—everything. Because there was never, ever, at any point in our lives or in human history, a perfect egg of goodness to shatter. Why would there be? For fourteen billion years of random, purposeless, unguided evolution, matter floated around, formed stars that lit up and were extinguished, and those stars eventually formed the material that formed you and me, and also lizards, and garbage dumps and concentration camps. Why would we expect any perfection?

Of course you might say, what's the harm in a little metaphor? The folks at, say, *Tikkun* magazine know enough to see your point. *Tikkun* is a leading progressive Jewish magazine, founded by a very talented rabbi named Michael Lerner. The magazine offers a lot of good social commentary, and the communities associated with it are involved with a lot of worthwhile social justice work. They got the title from a phrase, *tikkun olam*, a Hebrew phrase that means "repairing the world," which was brought into the mainstream Jewish lexicon a generation ago by Abraham Joshua Heschel and others. It was intended to be a way to unite modern Jews who had an extremely wide range of beliefs and knowledge about God, the Torah, and prayer—including many who had no beliefs or knowledge about such things. *Tikkun olam* was presented to them as a way to be Jewish authentically without having to buy into all that theological stuff. And of course there's something to be honored in any idea that seeks to unite a diverse group of people to do good for the world.

But the metaphor of *tikkun*, of repairing, is precisely wrong. And metaphors can be very important. Right now, for the sake of our environment and to combat our terrorism-supporting oil addiction, we could desperately use a zero-emissions car that could get unlimited mileage from sustainable energy sources. But if I told you we needed to *repair* that car, you'd look at me as if I had three heads. Such a car never existed, so how can we repair it? We've got to *build* it. And saying that the difference between the two approaches is mere semantics is like Sarah Palin suggesting that it makes no difference what caused global warming, we just have to figure out how to fix it. How are we supposed to figure out how to fix a serious problem if we're confused about how and why it started?

Ceasing to believe in God or religion becomes a truly meaningful, worthwhile position when it also means ceasing to live in the past. We move on. We focus not on who wronged us, who screwed us, but on what we can do, what we can build, how we can grow, to make our love lives or our political lives, or any other aspect of our lives, better.

I have a friend named Sarah who was diagnosed a few years ago with an illness called polycystic ovary syndrome. It's one of those illnesses that even today's best doctors are still struggling to figure out. Sarah went to the best specialists, but all they were able to tell her was that, without aggressive hormone treatment, this condition would certainly rob her of her ability to become pregnant and give birth. Well, if you met Sarah, you'd probably notice two things about her: first, she's a warm, extremely affectionate woman—she earned a master's in education because she loves children and has always wanted several of her own; and second, she's quite a determined individual. So when she had a nasty reaction to the hormone therapy the doctors suggested, she didn't accept their prognosis that she'd never have children of her own. Instead, she began researching alternative therapies. Sarah changed every aspect of her lifestyle. She learned to plan every bite of food according to what would be most wholesome and nutritious. She meditated every day and went for acupuncture regularly, exercised, and did more yoga than a Yogi. And she cured herself. After a year of all this, she was given a clean bill of health by a team of baffled specialists.

But that's not the point of this story. If it were, I could just be making it up. Or you could simply and rightly ask, what about all the people who work hard to cure themselves of various illnesses and fail?

The point is that after all the study and hard work Sarah put into getting healthy, and the self-discovery that went into it, she decided to start a nonprofit called Dharma Harvest on her family's farmland. The project teaches agricultural skills to underprivileged children, bringing them to the farm to work and then bringing the fruits of their labor—sustainable, organic, delicious local produce—into public schools. But one day Sarah was telling me about how she dreamed of adding a special garden to the property, where patients suffering from serious or debilitating illnesses could learn to heal themselves the way she did: through better nutrition, fitness, meditation, and all the rest. She was thinking of calling it a "Healing Garden," though for some reason she wasn't crazy about the name.

I loved the idea, but advised her to call it a "Growing Garden," not a "Healing Garden." Because many of her visitors might never heal, the way she did, but as long as we're alive, we're always growing. Sarah wanted this to be an honest, healthy place for anyone in her community. And if someone came to her with a fatal disease, wouldn't it be dishonest to suggest that she could heal that person? You're no miracle worker, I told her—if they're going to die, no amount of meditation or Reiki or acupuncture is going to stop it. So if all you're doing is healing, then for that person with a fatal disease, what you have to offer is worthless. And the thing is, everyone has a fatal disease—don't we all? It's called being human. We're all going to die. No one can do anything to change that. So if medicine and the good works we do are just about healing, then they're really little more than putting a little Band-Aid on a very big boo-boo. And what's so significant about that? Not much.

But, I said to Sarah, as long as we are alive we can all still *grow.* Even if we have only one year, one week, or even one *hour* left in our lives, we can still use that time constructively.

My father died without having said the words *I love you* to me. I know he loved me. He showed it by many of the things he did. But I missed hearing him say the words, and I'm pretty sure the reason he missed saying them is because he was scared to. Like so many men of his generation, he never had much of a male role model for how to express vulnerability and tenderness, so he must have felt like something awful would happen if he broke the mold of his own experience to communicate such things. I used to wish, for my own sake, that my dad had come around just before he died, and said those words to me. But part of coming to terms with those painful wishes, for me, was learning to also wish he'd done so for his own sake. So that he'd been able to feel the relief of letting his guard down and expressing love fully. Maybe it wouldn't have changed *everything* about our relationship, but it would have made a real, positive difference for both of us. We both could have grown together, even if he would have been almost done growing by then. Indeed, if you've ever been tempted to suspect that "we begin to die as soon as we are born," Humanism's answer is: so long as we continue to grow in some way, we are *living*, not dying.

None of us will ever fully recover or be healed from the human condition. No amount of repairing the world will ever actually fix it. But we can grow tremendously in our abilities to understand and to feel and give love and affection and empathy. We can build up those around us, and we can build a better world. That's Humanism in a nutshell—recognizing the difference between magic and reality, then bringing people together to help each other get on with the work of growing and building. The idea that we need to build—and not necessarily repair—can pervade every aspect of our lives. But two areas that need special attention are the new ethical problems—personal and societal—that have emerged because we have built a world radically different from anything our religious ancestors could ever have imagined. Modern life poses new challenges that no ancient religion anticipated, and no ancient religion can possibly have any better insights into these challenges than those of a reflective, serious Humanism.

Voluntary Simplicity

Cavemen had a lot of issues to deal with. But feng shui was not one of them. The complexity of life in today's world is a new problem that requires a new kind of solution. Older times were in some ways simpler times, yet we often forget that to go backward would mean not only an escape from BlackBerry Hell, but also a return to the racism, sexism, death in childbirth, and all the other terrible forms of suffering we've made progress in addressing. Yet we can't just keep going as we are. It would be foolish to ignore the fact that depression levels in this country are ten times higher than they were in 1900, partially because of the raised expectations that have accompanied our progress. And it would be worse than foolish by far if we forget that, to paraphrase an old Cold War saying, we may choose to fight World War Three with nuclear weapons, but if we do we'll certainly fight World War Four with sticks and stones.

In our Humanist community at Harvard, we recently began addressing this issue by offering seminars in Voluntary Simplicity. Voluntary Simplicity is a worldwide movement of people who recognize that a good life in the twenty-first century may value simplicity, but that doesn't mean we can or should simply cease all shopping, electricity use, or bathing. Rather, VS recognizes that when we become deliberate in our choices, we need fewer things to be happy. We can *choose* to focus more energy on those few things we do need, and let the rest drop away. The phrase may have been coined in 1936 by Richard Gregg, a student of Gandhi's who wrote:

> Voluntary Simplicity involves both inner and outer condition. It means singleness of purpose, sincerity and honesty within, as well as avoidance of exterior clutter, of many possessions irrelevant to the chief purpose of life. It means an ordering and guiding of our energy and our desires, a partial restraint in some directions in order to secure greater abundance of life in other directions. It involves a deliberate organization of life for a purpose. Of course, as different people have different purposes in life, what is relevant to the purpose of one person might not be relevant to the purpose of another . . . The degree of simplification is a matter for each individual to settle for himself.[18]

Especially in a world with increasing economic problems born of unsustainable commercial practices—an economy built almost entirely on tempting you to buy products you don't really need, with money you don't really have (and I say this as someone unashamed to call himself a capitalist! Or at least a "progressive capitalist")—the least we can do is pay serious, sustained, collective attention to certain VS practices like changing modes of transportation; looking for smaller-scale, more human-sized living and working environments; reducing, reusing, and recycling; shifting to a more sustainable diet; and lowering overall levels of consumption in favor of increasing levels of human interaction, cooperation, and mutual entertainment.

This is not about living a life of poverty, or being against progress. It's not a romanticization of the rural countryside. It's not about averting our eyes from the beautiful things and experiences that modern life can bring. As one leader in the Voluntary Simplicity movement has said, "Instead of a 'back to the land' movement, it is more accurate to describe this as a 'make the most of wherever you are' movement."[19]

Sustainability

Unfortunately, as important as it is to "make the most of where we are," it will not be enough. Humanist ethics demand even more, because human thoughtlessness has endangered our very lives. In addition to Voluntary Simplicity, Humanists and the nonreligious have no good option but to learn how to live their lives in a sustainable way. We cannot hide from global climate change, the proliferation of nuclear and chemical weapons, and the recent global financial crisis fueled by fear that we have massively overextended our entire economic system. Religious and nonreligious people alike have built a modern world upon habits that are simply not sustainable. And we are running out of time. As Al Gore reminds us, human survival depends on our doing something truly unprecedented: making a decision together—not just as a family, clan, tribe, city, nation, or bloc of nations, but as *the human species*. We must decide, all of us together, to survive.

Of course, not all nonreligious people have come to terms with the environmental situation. And Humanists who have recognized the dangers before us are not alone in doing so. Christians, even many Evangelicals, have begun to speak about stewardship of God's creation. But I believe that secular people, despite our relatively weak organizations, must step forward and take leadership on these issues because, as my former professor, the theologian Gordon Kaufman writes, "This [ecological crisis] is a different kind of issue than Christians (or any other humans) have ever faced, and continuing to worship a God thought of as the omnipotent savior from all the evils of life may even impair our ability to see clearly its depths and significance . . . [W]hat is now needed is a reordering of the whole of human life around the globe in an ecologically sustainable manner—something heretofore never contemplated by any of our great religious (or secular) traditions."[20]

Again, talk of returning to the way things were—to some sort of Garden of Eden mentality—does us no good here. Would we really want no roads, no toilets, no hot showers, no way to communicate with our loved ones when we're not with them, no heat or air-conditioning, no ability to safely cross oceans or even cross the United States? It would be criminally stupid to regret or vilify the scientists and innovators who created these new realities. But we cannot ever again afford to engage in this kind of discovery without *planning* every step of the way. We need, not divine guidance coming to us in a flash from the sky, but human wisdom, coaxed and cultivated methodically, purposefully. We now desperately need, not just city planners asking where to put this road, how long that stoplight should blink, or whether the park needs another bench or a water fountain. We need to plan better ways to use our waste and ways to draw power from renewable resources. We need to compete, as young people around the country have begun to do, over who can take the shorter shower and who can compost more comprehensively. For the first time since the days of Don Quixote, working with windmills may be among the highest forms of spiritual practice.

As a teenager going to high school in lower Manhattan, my favorite vista was where the World Trade Center met the park along the Hudson River. You could stand by the Hudson, watch it shine like the lyrics to "America the Beautiful," then turn around and see a large red Zen pagoda, a DNA-inspired sculpture garden and exotic imported trees, and finally look directly up at two gorgeous towers, reaching toward whatever God might reach back; impossible towers that ceased to be possible in September 2001. There was something about that combination of nature and human ingenuity that was especially awesome. And now it's gone.

But even before there were three disaster movies every season about blowing up the Statue of Liberty, there was the original classic *Planet of the Apes*. In it, Charlton Heston is a brave but bitter American astronaut who winds up on a mysterious planet he doesn't realize is Earth. It's two thousand years from now in a future when human beings have blown themselves to kingdom come. The planet is ruled by intelligent apes. The apes, afraid of Heston for many reasons, take their time in revealing to him where he is and how it came to be that his entire civilization was wiped out. When he discovers the truth, this astronaut who had gone into space to look for "something better than man" cries out:

> *"We finally really did it! . . . You maniacs! You blew it up! Oh, damn you! God damn you all to hell!"*

I wouldn't go so far as to ask all human beings to believe, as I do, that there is no heaven and no hell. But it is truly a sin to go on fixating on our petty little differences over territory, sexual morality, and other theological minutia while failing to acknowledge that there can be no greater ethical failure than allowing any of our differences—religious or secular—to make this scene one day truly come to pass.

CHAPTER FIVE

Pluralism: Can You Be Good *with* God?

How should we respond to this phenomenon that many call "the new atheism"? In *The God Delusion*, Richard Dawkins suggests that religious education for children may be a form of child abuse. In *God Is Not Great: How Religion Poisons Everything*, Christopher Hitchens picks up Dawkins's line of thinking and takes it at least a step further, offering ominously that the time has come to get to know the religious "enemy" and "to prepare to fight it." One would like to dismiss such a statement as a mere rhetorical flourish, but it should be noted that Hitchens was one of the most eloquent advocates for George W. Bush's all too real war in Iraq. Sam Harris has written that "Science Must Destroy Religion," and that perhaps even religious moderates—mere "failed extremists," he says in *The End of Faith*—should not be spared this destruction. Most recently the popular comedian and talk-show host Bill Maher has, if such a thing is possible, escalated the war of words even further by preaching in his film *Religulous* that "for humanity to live, religion must die."

Most Nonreligious People Are Not Antireligious!

It cannot be said often enough: Humanism stands for more than these incendiary statements. Let me also reiterate, regarding my previous chapter on Humanist ethics, that I am not claiming that every person who refers to him or herself as a Humanist always behaves ethically. And I'm not saying that religious people are always unethical. My point is merely that we Humanists have our ways and goals, and they're at least every bit as good as anyone else's. If you're prepared to say or think anything nice about anyone who disagrees with you, by this point you ought to have plenty of nice things to say about Humanism. If you're *not* prepared to say or believe anything positive about those who disagree with your theology, then we may indeed have to defeat you, but we will have many religious allies in doing so, as we saw in the 1800 and again in the 2008 American presidential elections.

There is no question that religious people have killed too many in the name of their God. But secular people have at times killed for their beliefs as well. Granted, almost all secular people (including the new atheists and their most passionate supporters) go about their beliefs peacefully today, and it would be immoral for me to try to make religious people feel better by making an equivalence, implying that we Humanists were out there killing just like the Taliban even today, when we are clearly not. Still, even if our percentages are better, there are still an enormous number of people who call themselves religious who are peaceful, open-minded, and not worthy of hate. Should we go to war with these people? Destroy them? Are they poison? Are they deluded?

We do not need to go to war with religion—not physically and not rhetorically either. But if not war, then what should the relationship between religious and nonreligious people be? There is a one-word answer: pluralism. This doesn't mean the end of all differences between us, or even competition among us. It means that competition should be, as in the Qur'anic turn of phrase, to "compete with one another in good works." (Qur'an 5:48)

I've been thinking about pluralism since I was too young to remember, because I grew up in Flushing, New York, "the most diverse neighborhood in the most diverse borough in the most diverse city on the planet," according to a recent *New York Times* op-ed.[1] Flushing was the site of one of the very first victories for interreligious acceptance in the New World. In 1662, the farmer John Bowne was banished from New Amsterdam for harboring Quaker meetings. He was not a Quaker himself but believed they deserved refuge from Peter Stuyvesant's intolerant regime. So Bowne set off to (old) Amsterdam immediately—not to defend himself, but to defend the Quakers. He won his case, and so did tolerance. As another citizen of the colony wrote even earlier in the century in a petition defending the Quakers, "We desire therefore in this case not to judge least we be judged, neither to condemn least we be condemned, but rather let every man stand and fall to his own master."[2] And more than three centuries later, the neighborhood was so teeming with religious, ethnic, and other diversity that when my friend Tari and I were assigned to go in front of our second-grade class and talk about our similarities and differences (as the teacher informed my parents later), we remembered to mention our blond hair and black hair, big ears (mine, of course) and small ears, his talent for sports and mine for art, but we completely forgot to mention that one of us was black and the other was white.

But it's not just that I'm from Flushing, and it's not just tolerance that I'm highlighting here. *Pluralism* is the heritage of my country, from its earliest days. President George Washington's single greatest act may have been his statement to the Touro Synagogue that "It is now no more that toleration is spoken of, as if it was by the indulgence of one class of people, that another enjoyed the exercise of their inherent national gifts. For happily the Government of the United States, which gives to bigotry no sanction, to persecution no assistance, requires only that they who live under its protection should demean themselves as good citizens, in giving it on all occasions their effectual support." Jews and all religious minorities were not only to be *tolerated* in America, but *fully equal.*

Still, neither the Qur'an nor John Bowne nor George Washington were in a position to answer the precise question that is before us today, which is, how should religious and nonreligious people relate to one another? But perhaps on this issue the New Testament and Martin Luther King Jr. have something to contribute when they suggest we should love our enemies.

I'm not talking about sitting and holding hands and singing "Kum Ba Yah," though that has, as Gustav Niebuhr eloquently points out in *Beyond Tolerance*, rather oddly become the put-down of choice when you want to make fun of people for trying to come together. After all, in the 1960s, he writes, people probably did hold hands while they sang it under intense pressure from southern police forces and Klansmen, and from there the song spilled over not to interfaith circles but to largely Christian youth groups and summer camps.[3]

A Meditation on Love and Pluralism

What I do mean is that we atheists and agnostics might offer nonviolent resistance to imposed religion in our lives—while loving our religious neighbors, offering them friendship and steadfastness even when they offer us spite. We might recognize that our humanity, in its fullest expression, can make some theists feel that their humanity—tethered as it is to belief that they are God's children—is called into question. We might say, "We will not hate you, no matter what. But we will not be anything other than ourselves, loud and proud." We might choose pluralism because, as Martin Luther King Jr. said in his 1967 Christmas Sermon on Peace, "we are caught in an inescapable network of mutuality, tied in a single garment of destiny. Whatever affects one directly, affects all indirectly."

Of course, there are real provocations. In most places in this world you are free to look a man in the eye and tell him you do not believe in his God, or in the platform of his party, or even in his right to marry the partner he holds dearest, and you may be considered disagreeable but you will still be seen as decent. But say that you do not believe in God at all, and despite whatever else you might add about the good things you do value, there are many who will consider you indecent and unfit. Nevertheless, there were real provocations for Martin Luther King and Gandhi too.

It's not that we should lie to ourselves and say we're their equals when we know that no one could measure up to their stature and their importance—perhaps not even them. But on the other hand, they were on to something, and we should be on to that something too. At a time, as Sarah Vowell writes, "when no one's typed the word 'nonviolence' since the typewriter,"[4] we should give serious thought to going into the most vehemently antiatheist, anti-Humanist parts of the world—parts of the Deep South of the United States, or the Arab Muslim world, for example—and starting a conversation (not to mention organizing community service projects) by just being ourselves, loudly and proudly, but in a loving, nonviolent way. Might it be dangerous if we happened to run into the one or two individuals in some town somewhere who can't physically countenance the presence of a proud Humanist or atheist? Yes, it could be. But if we're not willing to face even the slightest possibility of violence in order to spread our message, how can we expect change? I believe we have more than enough courage for the challenge.

In Dr. King's same Christmas Sermon on Peace, he went on to say that there are three words for *love* in the Greek New Testament. He did not mean we are obligated—or able—to love our enemies in the sense of having warm affection for them, but rather that we should cultivate *agape*: "understanding, creative, redemptive good will toward all men," and that by this kind of love, "We will not only win freedom for ourselves; we will so appeal to your heart and conscience that we will win you in the process, and our victory will be a double victory." But as beautiful as I find this language, I fear I'm going to get myself into trouble for using it. I sense that I'm putting myself in a no-win situation here, because fellow atheists may not be able to help but see this sentiment as excessively Christian, while most Christians will not think that by loving them as an enemy I do them any favors.

It *is* hard to like those who don't like Humanism, who don't like atheism, who would discriminate against me, who would be prejudiced. It's hard to like someone who wouldn't vote for me based on what I believe about the nature of the universe, and hard to like someone who would attempt to subvert the Constitution to proselytize to hundreds of thousands of soldiers on tax dollars. It's hard to like people who would not only violate my rights but would violate the rights of others I consider my friends and colleagues or just my fellow citizens and human beings, simply because they are gay or non-Christian or nontraditional in whatever way. But I don't want to hate the people who do this. And I don't want to turn my back on them and simply disengage from them, pretending they do not exist, that they are dead to me. I do not want to have to look at myself in the mirror and see a man who has inculcated that in himself which would see such people as entirely other, as entirely wrong and worthless and frightening, as enemies. I do not want to do to others that which I find hateful when done to me.

I know that I have only so much room for emotion in my heart and my mind on each day, in each year, in this life. If the emotion I choose to cultivate is hate, or indifference, or bitterness, then that's what I will become, and it doesn't matter if my so-called enemies have earned the scorn, because now I have given over to it too. So I want to feel another emotion toward them. It's not that I want to like them. It's not that I want to submit to their will. It's not that I want to be persuaded by them, that I want to be their best friend, or that I want any favors from them. But there is an emotion that I want to feel toward them. I don't simply want to enter into some sort of political negotiation with them. Political negotiations, when they are that and that alone, don't work.

For example, we all know what the right solution to the Israeli-Palestinian conflict would be. That's why I'm proud to be pro-Israel *and* pro-Palestine; it's why the most Zionistic thing anyone can do is work for a just resolution through the peace process in which two nations will live side by side, sharing Jerusalem as their capital. Because we all know it's either that or oblivion. The final status of each of those countries is already known, but nobody can move forward, because the emotions on the sides opposed to negotiations with their enemies are hotter than the emotions of those who want to negotiate. And the reason is that we are most fully ourselves when we admit that we are emotional beings, that we are defined by the ways we find though our behavior to express all these myriad emotions constantly bubbling beneath the surface of us as we try our fragile best to reason our way through the world.

And so yes, I want to love my enemies. Not in the strict sense of agape, of doing charity for them, and not even in Dr. King's sense of Christian love for them. Love has been around, I believe, longer than Christianity. And although I am the first to admit that Christianity is not going anywhere, love will be around longer than Christianity. I'm talking about love in the sense of caring. Caring: I think it's as good a definition as any for love. And that's what I want to offer to those who would be prejudiced and discriminate against me, that I care about you and indeed even love you anyway, and I too would rather die than hate you, though thank goodness many religious and secular martyrs alike have already come along and lived and worked and died so that I can hope not to have to chose between hatred and death.

I want to find loving respect for you and live to tell the tale—and see, as Dr. King proclaimed, that you too will be transformed and we will have a double victory.

But now that I've talked about my feelings on this issue, let's look at some much more concrete matters, starting with that definition I promised above of the religious pluralism that I'm calling for.

According to Dr. Eboo Patel, director of the Interfaith Youth Core,

> Religious pluralism is neither mere coexistence nor forced consensus. It is not a watered-down set of common beliefs that affirms the bland and obvious, nor a sparse tolerance that leaves in place ignorance and bias of the other. Instead, religious pluralism is "energetic engagement" that affirms the unique identity of each particular religious tradition and community, while recognizing that the well-being of each depends on the health of the whole. Religious pluralism celebrates diversity and welcomes religious voices into the public square, even as it recognizes the challenges of competing claims. Also, it recognizes that in a pluralistic democracy, competing claims must be translated into moral language that is understood by fellow citizens—believers and nonbelievers alike—who must be convinced of the benefits of what is proposed.[5]

This is a definition and a concept I want to affirm. But it needs to be understood that we have equal rights in all that pluralism entails, because this has not been immediately clear up to now, as we saw earlier in the quotes on secularism as the common enemy of religion. Would some atheists reject this concept of pluralism? Of course. But plenty of Christians reject it as well, and you'd hardly think of holding an interfaith meeting without Christians because of it.

There are three specific issues related to religious pluralism that need to be addressed in order for us to be good together, with or without God: religious literacy, interfaith cooperation, and the inclusiveness of religious pluralism.

Religious Literacy

"We have a major civic problem on our hands," says Boston University religion scholar Stephen Prothero. In his best-selling book, *Religious Literacy*, Prothero argues that religion must be taught in American public schools because our national illiteracy on the subject endangers our ability to address important social problems like terrorism and the need for greater investment in science research. Prothero's point is well taken, and he even takes on the profound misunderstanding that secularists are responsible for removing religion from public discussion. "In one of the great ironies of American religious history," Prothero writes, "it was the nation's most fervent people of faith who steered us down the road to religious illiteracy."[6]

I would like to support Prothero's message heartily, but his otherwise good book (along with many books like it) is in need of a revision when it comes to atheism and Humanism. Prothero devotes only a few short paragraphs to each of these subjects, presenting nontheism as a kind of footnote to a bigger, more important conversation about the various religious traditions in the United States. For example, he gets it right that Humanism is a belief that "human beings can get along just fine without God" but dismisses it as "more an epithet of the Religious Right than a self-designation."[7] This approach is common, but no longer adequate. The reason people do not designate themselves as Humanists is that Humanism, secularism, and atheism have been unstudied, underresearched, and otherwise ignored by everyone from scholars of religion to the popular media. Remember, one in five young people in America now considers him- or herself nonreligious. There are half a billion to a billion people around the world to reckon with. The many courses on comparative religion that Prothero's work is inspiring need full-fledged units on Humanism and atheism in their curriculum, not a mention of nonbelievers that you could miss if you coughed. Daniel Dennett gets at this same idea in his *Breaking the Spell*, providing an example of a supposedly hard-line, fire-breathing new atheist (actually, when you get to know Dennett in person his manner is just as much that of a scholarly Santa Claus as his looks) pushing to get "more education about religion into our schools, not less. We should teach our children creeds and customs, prohibitions and rituals, texts and music, and when we cover the history of religion, we should include both the positive . . . and the negative."[8]

And lest we think that religious people won't extend us the courtesy of teaching about Humanism and atheism, consider the case of Sheikh Hamza Yusuf, an American-born Muslim leader who, in a captivating keynote address at the 2007 national conference of the Interfaith Youth Core, told a room full of five hundred religious young people that they should all consider reading Richard Dawkins's *The God Delusion*—that it was a worthwhile book and one should always challenge oneself to understand all perspectives on religion. Only when I went up to Yusuf afterward to thank him for the magnanimous gesture did he privately and with a humble smile relate the story that Dawkins once publicly called him "palpably stupid." In fact, Hamza is palpably smart, and not only can we learn from his approach to religion, we atheists should learn to be religiously literate.

Interfaith Cooperation on the Big Issues

The major political and social challenges of the twenty-first century cannot be addressed by any one group alone. Dream if you wish about a time when religion will be no more. No one can stop you. But in the meantime, reason requires us to acknowledge that religion is here to stay, and we human beings may not be if we do not find the collective moral motivation to beat back climate change, rein in terrorism before it realizes its most destructive hopes, and prevent the erosion of our democracies as economies shift and hopes are dashed. My hope is that, on some of the issues listed below, Humanists can play a mediating role in the decades to come. We may not be the biggest or certainly the richest group, but if we keep in mind Margaret Mead's insight that a small group of thoughtful, committed citizens is the only thing that has ever changed the world, we may just become the most influential group in getting the world's religious communities to sit down together and work out successful plans and policies where they are most desperately needed.

Climate Change: Because Global Warming Doesn't Care What We Believe About God

E. O. Wilson, the two-time Pulitzer Prize–winning sociobiologist and committed Humanist, often called the "new Darwin," has been working tirelessly to assemble coalitions of Humanist scientists and conservative Christians to save "The Creation," regardless of their differing views on what created this world. At our 2007 New Humanism conference, I had the opportunity to address, via satellite, several thousand Evangelicals and others gathered to hear Wilson's plea for the natural world at Samford University (the "Ivy League of the Southern Baptist Convention") in Birmingham, Alabama. They seemed eager to hear from a conference of a thousand Humanists and atheists at Harvard, and to find common ground where possible. What struck me about the experience was that for every stop Wilson can make on that sort of tour, there are dozens of communities that need to be reached where he will not be able to go. We all need to be ambassadors for our community, and for the earth. I've told Ed only half-jokingly that if I could somehow be one of several Robins to his Batman in this cause, it would be an incredible honor.

Church-State Separation: Maintaining It Successfully Requires Compromise and Coalition

This is not the place for a long description of church-state separation, except to emphasize that Humanists and the vast majority of nonreligious people affirm the value and importance of secular law. Separation of church and state must be absolute; as we saw in Thomas Jefferson's vision, this benefits all people, religious or not. In a secular government, Humanists might have a unique perspective on any given moral or ethical issue. For example, given our belief that science is a much better method than revelation for determining the nature of reality, we are, if I might attempt a bit of an understatement, highly likely to support the teaching of evolution in public school science courses, and reject the teaching of "intelligent design" (as nonscientific) in those same courses.

But we Humanists do not seek to impose our view on the secular moral and legal systems. Rather, we see our views as no better and no worse than anyone else's when it comes to whether they should become secular law. We need to build consensus with other groups in order to find solutions that work for all people. And this is precisely what we've done in the past. To continue with the example of evolution, what most people on either side of the religious fence rarely stop to consider is that most religious people agree with the Humanist position, for their own reasons. The Catholic Church, the world's single largest religious denomination, has officially affirmed that evolution is real. So have all the mainline Protestant church denominations, most organized Jewish groups, and many more. Even a goodly number of Evangelicals have no problem with evolutionary theory and no patience for their brethren who clog the airwaves and the printing presses with their diatribes against Darwin. So, yes, we have plenty of allies on this issue. For more information on it, just look up the Reverend Barry Lynn, executive director of Americans United for the Separation of Church and State, who is an ordained Christian minister.

Arms reduction, poverty, and torture are among the many other issues on which most Humanists are progressive and thus will find countless millions of religious progressive allies with whom they can and must work. In fact I was very proud to join the National Religious Campaign Against Torture, and I recommend that atheists, agnostics, and the nonreligious support the NRCAT and its efforts to define torture as a moral issue and as an evil for which we must not stand. But I know that the NRCAT has been reluctant to publicly acknowledge that theists and atheists can share equally in denouncing torture. We must all work together in the promotion of such true moral causes.

Full Inclusion in Interfaith

Before we can understand why Humanists and atheists must be invited, and must choose to participate, in interfaith activities around the world, we need to ask what *interfaith* means. Gustav Niebuhr, in his book *Beyond Tolerance*, writes: "At its heart, it's a grassroots educational process in which the goal is to gain knowledge about individuals and their beliefs in a way that lessens fear. It is a new activity in the world, an entirely new phenomenon in our history. It is a social good, a basis for hope, and a tendency that ought to be nurtured and cultivated."[9]

We know that interfaith is the model, if for no other reason than that one side destroying the other is not. And fortunately, the way to promote interfaith work is not by promoting belief in what Allen Ginsberg called "Allee Samee," the lowest common denominator bringing people together to spout platitudes back and forth. "Humanists and Muslims are really the same because . . ." is never a good sentence, no matter how you choose to end it, no matter how decent and noble your intentions. Let's allow people, as Jonathan Sacks titles his book, the *Dignity of Difference*.

Eboo Patel and his very talented staff at the Interfaith Youth Core are living out their belief in pluralism in an impressive way. The IFYC has gone out of its way to include Humanists and atheists in recent years, though he admits he did not start the organization with that intention. Mainly, Patel—a progressive, religious Muslim—wanted to bring together young leaders from many backgrounds, to be an alternative to the corps of leaders being trained by Al Qaeda and other extremist religious organizations. These young people, now by the tens of thousands each year, do community service and educational projects together, and if you just watch some of the IFYC's videos or get involved with them locally, you may agree it's remarkable work. But Eboo originally assumed it would just be for religious people, and to his credit he's been willing to learn: "In the great American pragmatist tradition, it just happened," he told me, that atheists and Humanists became one of his core constituencies. But as my mom used to say, "Nothing *just happens.*" The IFYC became truly pluralistic because consistently, over the course of several years, one in five young people attending their events was nonreligious, and they were open enough to recognize this and honor it.

Full Inclusion of Humanists and Atheists in Public Interfaith Ceremonies

In a historic step, the Democratic Party unveiled its 2008 convention with an "Interfaith Gathering of Clergy." It had learned well from leaders of the religious left: innovative thinkers like the Evangelical Jim Wallis and the rabbi Michael Lerner, who preach that poverty, the environment, and education are deeply moral and spiritual issues and that Democrats must not abandon religious voters to those who reveal a narrow, puritanical, and too often hypocritical obsession with sex. Raise your hand if you would have predicted just seven years earlier, on 9/11—not to mention seventy years ago—that a major American party would begin a convention honoring a black man by honoring a Muslim woman cleric standing alongside an Orthodox rabbi and many other representatives of America's diverse communities. I don't care how much of a secularist you are—if you couldn't find something heartwarming about this event signaling our nation's expanded horizons, consider having your ticker checked.

However, historic moments like these will ultimately become still more examples of prejudice if we open our doors and our podiums to new religious groups without acknowledging the equality of Humanists and their values as well. Will "faith" end up as nothing more than a reason for division between religious and secular Americans—a cheap euphemism for belief in God, miracles, and the supernatural, as opposed to reason, empirical evidence, and this-worldly ethics? We can do better.

This is how I think we can make inclusion work:

Don't ask, "Can you be good without God?"

Do ask why we are motivated to be good, or to work with you.

Don't proselytize to atheists in an interfaith context.

Do reach out specifically to atheist, secular, and Humanist groups and solicit their *participation*.

Don't advertise interfaith events as for the religious only or as a way for everyone to unite, despite theological differences, around belief in God.

Do advertise as religiously pluralistic, including all religions as well as atheists, agnostics, Humanists, and the nonreligious. You don't have to mention all those words if you don't have time or space in a very short advertisement—pick whichever one or two you like. But if you have the space to mention several religions by name, consider lengthening your description of us as well. You might also say you welcome "all religious and ethical perspectives," or as it is often phrased in Europe, "all religions and *lifestances*."[10]

Here are my three suggestions (as opposed to commandments) to help recruit stronger participation from Humanists and atheists for an interfaith group:

1. Use inclusive language: In addition to including us on your usual flyers, posters, or recruiting e-mails as above, try a special poster or e-mail emphasizing that *interfaith* includes the nonreligious too. If you're overrun with friendly heathens, you can put a stop to this practice, but if it gets you a number of nonreligious participants roughly equal to the number of Jews, Muslims, or Methodists you've been recruiting, you're doing well.

2. Include us in programs: Does your group make visits to various churches and other houses of worship so that members can learn about each other's practices? Do you bring in speakers to talk about various traditions? If so, reach out at least once a year to local, regional, or national Humanist or secular groups who might provide you with a speaker or connect you with a local affiliate to visit. If you're the sort of organization that regularly asks clergy of different persuasions to offer prayers or invocations, invite a Humanist chaplain, rabbi, minister, or lay leader to do one (I'm asked to do these all the time and very much enjoy the opportunities to meet different communities used to hearing from ministers, priests, imams, and the like). You'll feel good having stood up for your own principles of pluralism and inclusion. And if you do get the chance to arrange a visit, but you're not sure what to do on the outing, try arranging a mutual community service project—preferably one where you can do some on-the-ground work together and share stories while you work.

3. Learn and teach about us: As a religious person, you will be a truly great ambassador for the interfaith movement when you

can gently but confidently educate nonreligious people about their own tradition. When you encounter someone who says, "No thanks, that sounds cool and all but I'm not religious, so it's not for me," ask what he or she means by *nonreligious*. If they're atheists, secularists, or Humanists, ask if they've considered learning more about the ideas behind that identification. You might talk a little bit about how Humanism and atheism go back to ancient times, and how they represent a lot of people around the world. Then direct them to one or more of the organizational Web sites listed in the appendix of this book, and encourage them to contact one. It will do you no harm to encourage someone who has already made a decision not to be religious to deepen his or her commitment, and those of us working in the Humanist movement will be impressed by your generosity and eternally grateful—or at least, grateful for a very, very long time.

Good With and Without God: Humanists in an Interfaith Nation

When Lori Lipman Brown, former director of the Secular Coalition for America (the congressional lobbying office for Humanists, atheists, and the nonreligious) was running for reelection to the Nevada State Senate in 1994, she was attacked by her opponent, Kathy Augustine, for not participating in some explicitly Christian prayers held in the Nevada State Senate chambers. Augustine and some of the Republican leaders of the Nevada Senate put together a public relations campaign to smear Brown as not only non-Christian but unpatriotic, going to reporters with the false accusation that Brown "actively opposed prayer and refused to participate in the Pledge of Allegiance in legislative sessions."[11] Ads were taken out in local newspapers contrasting the supposedly unpatriotic Brown with Augustine as a self-professed "active church member and proud to salute our nation's flag." The accusations stuck and Brown lost. But years later, Augustine was forced to admit to willfully violating state ethics laws. She and the former state senate majority and assistant majority leader eventually sent Brown signed retractions of their accusations, acknowledging that they had misled the public. As Augustine put it, Brown "had never actually done anything to my knowledge which showed anything but the utmost respect for our flag and for the veterans of our nation." But the damage had already been done.

Brown thought back to that time with deep satisfaction in 2007, when on behalf of the Secular Coalition for America she accepted membership into the Leadership Conference on Civil Rights—the largest civil rights lobbying organization in the nation. The LCCR—founded by African American labor and civil rights leader—and Humanist—A. Phillip Randolph, lobbied for and won the passage of the Civil Rights Act of 1957, the Civil Rights Act of 1964, the Voting Rights Act of 1965, the Fair Housing Act of 1968, and also helped to organize one of the defining events of the twentieth century—the 1963 March on Washington. Today it fights not only for African Americans but for all "persons of color, women, children, labor unions, individuals with disabilities, older Americans, major religious groups, gays and lesbians," and more. And with the addition of the SCA, from now on it will also defend the right of Humanist candidates to stand on an equal footing with Christians, certifying that discrimination against the nonreligious can indeed be considered a civil rights issue.[12] Perhaps thanks to Brown and many others like her, a number of talented, civic-minded young people reading this book today will one day successfully seek political office as open and proud Humanists, carried to victory based on merit and determination and supported by religious and nonreligious voters alike.

As with other civil rights struggles, sometimes we can make progress toward our ultimate goals in unexpected ways. This was certainly the case in Philadelphia in the summer of 2008, when a clash of roadside billboards gave way to a moment of interfaith understanding and cooperation.

Earlier that year, a group called the Philadelphia Coalition of Reason (PhillyCoR) had come together for the purpose of uniting the city's numerous Humanist, secular, and atheist groups toward the common goal of raising the profile of the nonreligious in the area. To that end, the group rented a billboard along a busy section of I-95 to display the message "Don't believe in God? You are not alone." But it happened that just up the Interstate, a large local church, the Light Houses of Oxford Valley, had taken out a billboard to display a similar sign—both had a bright blue sky with floating clouds in the background—to encourage drivers to "Experience God." When members of the ostensibly rival groups became aware of each other's efforts, they braced for a long (and potentially very expensive, if not bloody) advertising war. But then they came up with a better idea.

In a story for the *Philadelphia City Paper* entitled "Religious Smackdown!! (Not Really)," local reporter Boyce Upholt playfully described how the church members discovered their competition and realized that their God might need to "up his game."[13] On the church's Web site, Pastor Bob Jones offered this challenge: "I am not asking you to believe, but simply open your eyes and minds and see if there is something more." Reading the blog was my friend and fellow Humanist activist Martha Knox, coordinator of the PhillyCoR group. Martha reached out to Light Houses and offered, instead of more slogans, for PhillyCoR to join them in a day of joint community service. "We want those who disagree with us to understand that we share the same secular values," Martha said. "Charity is a secular, human value, not [only] a religious one."

As Upholt described it,

> And so, on Saturday I joined a crowd of Christians and atheists, 20-odd of each and all wearing T-shirts to mark their allegiance, at the Philabundance warehouse in North Philadelphia.
>
> Things began quietly. The groups mingled in the parking lot, waiting for someone to take charge. When I was discovered as a neutral party, I was assigned to take a group photo. Atheists and Christians clustered around a picnic table, patterned together awkwardly like boys and girls in an elementary class picture. It was hard to tell them apart: Both sides consisted of nice families with children and polite young adults. Even their T-shirts looked the same, white with subtle blue logos across the chest. Only the details gave a few away: a "Spirit in the Sky" ringtone on one side; Knox's edgily short hair on the other.

In the end the groups got along very well as they worked together to pack personal and medical supplies for homeless shelters. No one was rude, and in particular the teens on both sides found they had a lot in common. The reporter, clearly sensing he had a great story on his hands, seemed to

Knox to be waiting for someone to break into a debate about theology, but it was not to be. "What's right in your heart is right in your heart," said one atheist mother. And "People are in different places in life," said Pastor Heidi Butterworth. "Hey, you guys are other people in the community. We love you and God loves you. It's simple."

Humanists and atheists have learned from the experience to focus on deeds, and PhillyCoR has continued to get together for a community service outing every month since. Occasionally we all need to remember that we can get beyond arguing over whether we can be good with or without a God, and simply be good, together.

CHAPTER SIX

Good Without God in Community: The Heart of Humanism

It was only a matter of time, and it was bound to happen first in America.

Humanism today can be categorized as a movement, a philosophy of life or worldview, or perhaps you prefer as I do the European term *lifestance*. It's the melding of a comprehensive philosophy with a world tradition and deeply practical ethical and social commitment. Regardless, for better or worse or both, modern, organized Humanism began, in the minds of its founders, as nothing more nor less than a religion without a God.

By the early twentieth century, several revolutions had for some time been competing to outdo one another—scientific, technological, and industrial—not to mention the aftershocks of widespread democratization still coming from the American and French Revolutions: emancipation, suffrage, socialism, and global capitalism. Biblical criticism, anthropology, psychology, and sociology were rewriting much of our religious past. For those attempting to educate themselves according to the latest body of knowledge, whole limbs of the past had simply been blown apart. And if America has always been, as G. K. Chesterton called it, a "nation with the soul of a church," all these changes were having a unique effect on its soul.

It's easy to forget, today, that the first English arrivals at the Massachusetts Bay Colony in the 1600s—the founders of Harvard College and its surroundings as America's original "City on a Hill"—had been persecuted in Europe for being *too* religious. But there is no better word than *fundamentalist* in our language today to describe the theological and social style that those early colonists wanted their religious freedom in order to pursue. Of course, there were liberal dissenters almost right away, and within just three generations, many of the very clerics Harvard was established to train were preaching early versions of the heresy of "Unitarianism"—Jesus as a human being and the human spirit as divine. In his own day, Thomas Jefferson predicted that Unitarianism would sweep the South.

In fact, Unitarianism and other liberal versions of Christianity did sweep beyond their famous dominance of Harvard to much of the Northeast United States, and headed west as well. By the turn of the century there were thousands of churches with ministers and congregations at least partially open to evolution and the host of other new ideas emerging from the natural and social sciences. It was in one of these Unitarian churches in Spokane, Washington, in approximately 1915, that a bright and charismatic young minister named John Dietrich (1878–1957) first heard the word *humanistic* used to refer to the radical new set of ideas he'd been formulating. Just four years earlier, Dietrich had been tried for heresy and removed from the pulpit of a Reformed Church in his native Pennsylvania, despite having recently doubled its membership. In his move to Washington State he was able to increase church attendance from sixty to over fifteen hundred while further developing a doctrine that was liberal even beyond incorporating evolution and denying the Bible's infallibility and Jesus's virgin birth or divinity.

Dietrich had been looking for new language to describe his vision for a church community that would leave behind any notion of God and the supernatural, instead gathering regularly to celebrate and deepen human knowledge and ethics, by means of science and the humanities. He liked that the word *Humanism* echoed with connection to the renaissance Humanists in Europe, who had pushed Christianity away from rigid dogma about the mind of God in heaven toward an embrace of this life for its own sake. And so the outspoken leader adopted Humanism as the name for his religion, and began to preach it.

Though a great orator, Dietrich never wrote a book and retired from the pulpit while still in his prime, having made only a limited personal contribution to Humanism beyond the congregations he served. However, in his use of the word *Humanism* to describe his new ideology, he strongly influenced two younger and like-minded Unitarian ministers, Curtis Reese and Charles Potter, who had been calling their new ideologies a "religion of Democracy" and "Personalism," respectively. Dietrich is often called the "Father of Religious Humanism" because he was the first to crystallize the idea of an American church without a God, but Reese and Potter did more to organize around and promulgate the idea nationally. The three became acquainted with one another in the late 1910s, and by the early 1930s they had gathered around them a circle of supporters that included many prominent academics and intellectuals and numerous other Unitarian clergymen. In 1941 many from this initial circle were among the founders of the American Humanist Association, still the leading body for Humanist activism in the United States today.

Even then there were plenty of ideological divisions, as happens any time a group comes together for any purpose—the old joke "two Jews, three opinions" could apply to Humanists (or anyone else) just as well. Early on there were those who objected (as a majority of Humanists now most likely would) that Humanism, godless and without dogmatic authority as it is, need not be called a religion. And even in the early years there were battles over style and pluralism: historian of Humanism Mason Olds recounts that at a Humanist meeting in the 1950s one attendee said to a more aggressively antitheistic colleague, "One does not need to be hateful to be an atheist."[1]

More important than these internal disagreements, however, were the intense ideological struggles the first generations of Humanists fought, alongside liberal religious allies, to defend civil liberties, women's rights, science, and progressive values of every kind. The list of twentieth-century Humanist greats as demonstrated by the American Humanist Association's list of "Humanists of the Year" reads like a diverse Who's Who of the progressive heroes of the age: Margaret Sanger, Abraham Maslow, Carl Rogers, Ted Turner, Isaac Asimov, Alice Walker, Kurt Vonnegut, John Kenneth Galbraith, Carl Sagan, and Betty Friedan, to name only a few. Perhaps it was this success in attracting the affiliation of the era's leading minds, along with the passionate commitment of many Humanists to pushing back against social conservatism, that evoked the ire of the religious right to a degree disproportionate to the level of success Humanists have ever had, up to now, in recruiting members or funds. Over the years, right-wing Christians have lustily poured energy and resources into demonizing Humanism, culminating in the 1970s and 1980s with spirited accusations from most prominent leaders of the "Moral Majority" (Jerry Falwell, Pat Robertson, Tim LaHaye, and others); for example, that "the liberals and Humanists are slowly 'sneaking in' perverted and antimoral sex-education materials among public school systems,"[2] and that it is "time that 175 million or more pro-Americans in this country go to the polls and vote out of office the 600 Humanists whose socialistic viewpoints misrepresent them."[3]

Indeed, even past the midpoint of the twentieth century, Humanism seemed to be a force on the rise in world affairs—a great story unfolding. Yes, the most influential social philosophers had long reached consensus that a reason-based naturalism was the only dignified way to the good life. However, Humanism's influence extended far beyond the elite orbit of salons and cafés; the question "Is God Dead?" made it all the way to the cover of *Time* magazine on April 8, 1966.

As empires waned and crumbled, new nations emerged led by Humanist secularists determined to enforce the separation of state and religion: Turkey, India, Pakistan, and even, largely, Israel. The United Nations had been founded to bridge the gaps between nations and cultures, and accordingly it had prepared a bold and (not accidentally) boldly secular Universal Declaration of Human Rights. The United Nations Educational, Scientific and Cultural Organization (UNESCO) was chartered and founded by devoted Humanist activist Julian Huxley and held prestigious conferences of experts on subjects like "Humanism and Education in East and West," not merely in Paris and London but in New Delhi and Abu Dhabi.[4]

In the social sciences, the "secularization thesis" was gaining more and more credibility—intellectuals tended to see the world as becoming progressively less religious, and the day when all religion would be extinct seemed not far away to a majority of mainstream scholars. This trend was typified by the smashing success of Harvey Cox's 1966 landmark book, *The Secular City*.

In sum, a new organized movement with great potential seemed to be dawning. The Humanist Manifesto, a cooperative product of thirty-four leading American religious and philosophical leaders including Dietrich, Potter, and Reese, was published in 1933, claiming, "THE TIME HAS COME for widespread recognition of the radical changes in religious beliefs throughout the modern world." Clearly all this gave reason to believe that for perhaps many millions of people, the time for Humanism had come.

Today, of course, respectable social scientists proclaim not the death of God but the collapse of the secularization thesis. On university campuses, the trendy talks on religion have long since shifted toward addressing a "global resurgence of religion"[5] as demonstrated by 9/11, the worldwide rise of Pentacostalism, the deadly influence of Jewish and Muslim extremists on peace negotiations between secular Jews and Arabs in Israel and Palestine, the pesky tenacity of the American religious right, the Danish cartoon crisis, and the dozens of other global flash points of religiously fueled tension and armed conflict—Kashmir, Tibet, Darfur, Bosnia, Sri Lanka, Rwanda, and so many more.

What Happened?

Many factors have contributed to Humanism's failure, thus far, to fulfill expectations. To some extent the movement may have been damned by its own success in having an indirect influence on broader society. Over the past century, large percentages of the believers in all the world's religious traditions have moved closer to beliefs that one hundred years ago were common almost only among atheists and agnostics—acceptance of evolution, and a focus on this-worldly ethics as opposed to supernatural narratives.

Still, there have been other problems. Too many Humanist, atheist, and secularist thinkers and leaders have placed too much emphasis on the secularization thesis—essentially, the mistaken belief that "history" would do their work for them. As Oxford scholar and critic of atheism Alister McGrath points out, Humanist and atheist critics of religion have often been blind to the fact that theism is a "moving target."[6] Ironically, *because* religion is the creation of human beings and not a God, and human beings are nothing if not adaptive and adaptable, religious institutions have always found new ways to thrive in response to criticism, rather than merely wilting away at the first sign of withering theological critique. And for every atheist who has imagined that religion would one day simply drop away like a husk from the kernel of atheism, the cause of Humanism has been deprived of what Martin Luther King Jr. called the "fierce urgency of now" by which the difficult work of institution building must be done.

We should also acknowledge that Humanism and atheism have been negatively affected in the United States by a perceived association with socialism and communism. Senator Joseph McCarthy's hateful minions succeeded in making the "godless Communist" the great American bugaboo of the years following World War II, not only by exploiting the Cold War, but also by using godlessness as a wedge against America's increased religious diversity. As reflected in Will Herberg's classic work, *Protestant, Catholic, Jew,* belief in God made a convenient unifying theme once Protestantism could no longer stand as the test of true Americanness.

But then it is also a fact that some Humanist and atheist leaders, in the early going, did see socialism and even communism as an expression of the highest human ideals. During the lean and frightening early years of the Great Depression this was perhaps understandable, but unfortunately there were also those Humanists who held their faith in one or another school of Marxism until the painful facts of the Stalinist and Maoist regimes were impossible to deny.

Still, the era of support for Marxism among Humanists is as dead and gone as the time when many American Christians and Jews used to feel such sympathies as well. Indeed, all the above issues are essentially water under the bridge. Not a single one of them is of even the slightest concern to the countless millions of young people born around the world in recent years who will undoubtedly grow up to be atheists, agnostics, and nonreligious. The vast majority of these people—you may be among them—couldn't care a fig for the *organized movement* of Humanism, because they've never heard of it or because, as it currently exists, it is not relevant to their lives. Why not? Because up to now, the single biggest weakness of modern, organized atheism and Humanism has not been the religious right or radical Islam or the secularization thesis or communism. It has been the movement's own tendency to focus on religious beliefs, when the key to understanding religion lies not in belief at all but in practice—in what people *do*, not just what they think.

Life in general, and religion in particular, is not as simple as deciding, "I believe" or "I'm an atheist." Especially now that God can mean anything you want it to mean, believer and atheist have become such broad categories, such blunt and imprecise linguistic instruments, that sometimes the only thing they really communicate is the kind of polarizing, talking-point divisiveness of which bad partisan political debates are made. Your relationship with religion, whether you're religious or not, is about more than which God you profess to worship or deny when asked in some census or on some cable TV talk show. It's about how you live life every day, how you respond to a thousand situations that are impossible to fully predict or prepare for.

In short: we've successfully responded to the head of religion, but not to the heart of religion. And not coincidentally, we've often produced a very heady atheism. But I believe in the heart of Humanism.

What happens when you have a broken heart? Not just romantically, but anytime you lose someone, something you want, need, and crave. Anytime your hopes and expectations are dashed. It feels so unfair, it hurts so much, and you want to cry out, or at least make a little request. It doesn't even need to be to an orthodox God for many. It could be, *Starlight, star bright, first star I see tonight, I wish I may, I wish I might . . .* But you want to fall back on the ritual. You want to fall back on something.

To Sing and to Build

What the brief thumbnail sketch of modern Humanist history above comes down to is this: Humanists, atheists, and secularists have learned to do two things extremely well over the past century—to speak and to debate. We have articulated positions on a number of crucial issues, and defended those positions against all manner of often unfair attacks. But now we need to sing and to build. We need to acknowledge that as nonreligious people, we may not need God or miracles, but we are human and we do need the experiential things—the heart—that religion provides: some form of ritual, culture, and community.

Actually, as the first Humanist minister, Dietrich was also among the first Humanists to explore the idea of rituals without God. And the model he created has, amazingly, stuck around for almost a century as the most common one seen at meetings of Humanists and atheists in the United States. Dietrich got rid of whatever parts of the church service he disagreed with ideologically—namely, the whole church service—and kept the sermon. He saw church as an opportunity to provide people with continuing education. Perhaps this was a good idea in 1915 when most people did not have access to many other sources of new information or entertainment. But now we can get lectures anywhere. And I can tell you that, in a university community in particular, lectures are important but not enough.

While I was working on this chapter, a nineteen-year-old Harvard student named Peter died on a beautiful sunny fall Saturday morning. He was competing in the annual River Run, a race on the Charles River. The cause was a totally unexpected heart complication. One minute, he was a brilliant young man with a bright future—the pride of his family. The next minute, he was in cardiac arrest. As he doubled over, the dozens of students running alongside him stopped only gradually, not realizing what was happening.

An ambulance was called and the waiting began, as paramedics attempted to revive Peter. At this point, some of his friends and dorm mates gathered into a circle. Someone called them together to pray. They put their arms around each other. None of them even knew what Peter's religious beliefs or affiliations were. But the student who gathered them started in with a Christian prayer: "Jesus, please help Peter at this moment, be with him and with his family . . ." One of my students—Kelly, a second-generation atheist—was in the circle. At first, hearing Jesus's name invoked, she tensed up and felt defensive. But she continued to listen, and eventually found herself comforted by the ritual, by the moment of close companionship it allowed. She did not, could not literally believe in the words being said: that Jesus hears and heals. And indeed, none of the prayers said that day by Peter's family and friends were able to bring him back to life. But Kelly believed in the need to offer steady words together at that moment, if not for a miracle then for the sake of those standing by in shock. And she would not have known what to say on her own. She too was only nineteen, and unlike the Christian young man leading the circle, she'd not been brought up to participate in, let alone lead any kind of ritual. When she told me about all this a few days later, Kelly was feeling conflicted: glad to have been there for the prayer, but guilty for taking comfort in an act she did not believe in or endorse.

Beyond ritual, what is the role of culture more broadly in religion and Humanism? How should we understand the relationship between belief in God and religious affiliation? We've seen how many good arguments there are against belief in God. But remove religious affiliation, and for most people you also strip away their sense of connection to their unique ancestry, heritage, memory, and identity. Is the sacrifice worthwhile?

As a graduate student I had the opportunity to study with the late Samuel Huntington for a semester, and I used to debate him vigorously and publicly (to some of my shyer or more conflict-avoidant classmates' chagrin) about his "Clash of Civilizations" thesis. Huntington—in his book *Who Are We?*—pointed out that American identity used to comprise four elements: creed, culture, race, and ethnicity. He argued that the decline of ethnicity and race as markers of national identity, especially among whites, highlighted the weakness of creed or philosophy as the sole component of identity and directed Americans to refocus on culture:

> at the end of the twentieth century the Creed was the principal source of national identity for most Americans. Two factors enhanced its importance. First, as ethnicity and race lost salience and Anglo-Protestant culture came under serious attack, the Creed was left as the only unchallenged survivor of the four major historical components of American identity. Second, Creed had acquired renewed status, comparable to what it had in the Revolution, as the defining characteristic distinguishing America from the ideologies of its German, Japanese, and Soviet enemies. Hence many Americans came to believe that America could be multiracial, multiethnic, and lack any cultural core, and yet still be a coherent nation with its identity defined solely by the Creed. Is this, however, really the case? Can a nation be defined only by a political ideology? Several considerations suggest the answer is no. A creed alone does not a nation make.[7]

Culture is not only relevant to Americans of Christian background, Theologian Alister McGrath, in his recent work *The Twilight of Atheism* (whose title is obviously no more perfect prophecy than Sam Harris's *The End of Faith*), argues that religion has become almost synonymous with culture for certain groups in Western society, citing immigrant groups in particular:

> The role of religion in creating and sustaining communal identity has been known for some considerable time, and has become increasingly important since about 1965. One of the most obvious indicators of the ongoing importance of religion is the well-documented tendency of immigrant communities to define themselves in religious terms . . . [Immigrant] communities that have arisen within British cities self-define using religious (rather than national) parameters, *with places of worship acting as community centers.* The British media have learned not to speak of "Indian" communities in Britain, but of Sikh, Hindu, and Muslim communities, and to expect the identities of these communities to be focused on the local gudwara, temple, or mosque. A similar pattern is found in France . . . with the mosques of Paris and Marseilles sustaining the identity of France's five million Muslims [emphasis added].[8]

As a non-Westerner you have a hard time immigrating to the United States or Europe—I got a little taste of this in learning about my mother's journey as a refugee from Cuba. You know you cannot import too much patriotism for your former country—you have to make a good-faith effort at allegiance to your new host nation, even if the odds seem stacked against your eventual acceptance there. And so your house of worship can become simply a house of community and identity, a place where you can go for relief from the often relentless and unrealistic pressure to conform to your new surroundings. Is it really putting our best foot forward to tell people in this position that their religion is just a bunch of wicked nonsense?

In addition to ritual and culture, religion provides the human bonds of community that most people do not associate with Humanism or atheism. McGrath describes the situation regarding American Christianity:

> Christian churches have long been the centers of community life in the West. The more entrepreneurial of American churches have recently begun to develop this role further, seeing the church as an oasis of communal stability in a rapidly changing culture. [With] Radical and innovative approaches now being adopted in Christian worship and life . . . the success of [these churches] can be attributed to their recognition of the importance of creating a sense of community identity. People want to belong, not just believe. Such churches see themselves as "islands in the stream," offering safety and community to travelers on the journey of life. *Identity is about belonging somewhere. And the community churches see themselves as providing a place where members belong* [emphasis added].[9]

All people, not just Christians, want to belong to something. As much as we've been taught the sometimes very real downside of institutional affiliation, we have not changed human nature. No matter how much we value our inventiveness and our ability to think for ourselves—and we should value them greatly—we have not yet invented the man who is an island. And indeed, even we nonconformists should shudder at the possibility of such an invention as much as Dr. Frankenstein learned to shudder at his own monster.

The good news is that there are good alternatives, without God, to religious ritual, cultural identity, and community. We need only discover and develop them. Here is how.

Do Humanists Pray?

Prayer is one of the most important aspects of religion, and we need to offer meaningful alternatives to it, not simply dismiss it as silly. To understand why, think about the serious message underneath the old wisecrack that there are no atheists in a foxhole—even though the joke itself is untrue and not really all that funny either. The message is that in a foxhole, even the soldier who is least religious under normal circumstances can't resist the urge to pray to an unseen God: help me, protect me, save me from the cruel luck of bullets and bombs. And of course it's also a metaphor—there are no atheists when the chips are down, when you're in danger, whenever you want something so badly that all you can do, it seems, is close your eyes and pray for it.

Every human being learns to pray as a response to fear—maybe even before we learn to talk. We do not consciously remember this training when we are adults. But as tiny, helpless creatures, we do not have the ability to form words or even thoughts such as "I am hungry," "Hold me," or "I have to go . . ." Yet we cannot meet these needs on our own—with the exception of the last—and if we do not find help in meeting them, we will rather quickly die.

So one of the first things we learn is to cry out to those seemingly omnipotent beings surrounding us—our parents, or whatever other biological or surrogate family we have—and in the rawest, most primitive form of communication, we cry a *waahhh* that means "Help me!" Eventually, we discover that certain ways of crying out make help more and less likely, so we begin to modify and regulate our requests. But we will never be sure we'll receive the help we crave. This set of emotions, once we gain language and some theological training, becomes formal petitionary prayer.

As adults, our language grows more sophisticated. But so do our problems. And though we learn to use reason and science to solve some of the challenges we encounter, each success only brings new uncertainties. The child inside, crying out against uncertainty and fear, never entirely leaves us. This is why prayer is one of the most versatile tools human beings have ever designed for coping with stress and promoting feelings of empowerment. It is something we can always do to collect ourselves, even when alone, stranded on the proverbial desert island, with reason to fear that without all our wits focused, we won't survive. When we have almost nothing, if we have our consciousness, then prayer is an option. It can be based on brilliantly rhythmic poetic compositions—the Jewish mourners' Kaddish possesses a meter many Shakespearean sonnets would envy—yet prayer doesn't require a fixed text, or an appointed time. You can pray privately, while being watched, even while pretending to do something else. When you're frazzled and need to be soothed, prayer can be soft and slow. When you're depressed and need energy, it can burst forth with the full force of a Christian heavy metal band.

As the neuroscientist Andrew Newberg explains, regular ritual participation creates "resonance patterns" in the brain, making mystical experiences that shut off the "self" more likely, by confusing the parts of our brain that track our physical boundaries and map the space around us.[10] And when it isn't shutting off the self, religious worship can help people focus on solving difficult problems. In a moment of crisis, the act of kneeling, lowering the head, and whispering "Dear God, I need you" may seem helpful only insofar as it provides a relationship to the deity or divine intervention. But it actually provides an opportunity to collect oneself and marshal internal resources that might otherwise go unnoticed or untapped.

In short, when we talk about prayer, we're really talking about the equivalent of a highly versatile, always available, perfectly legal, free, non-physically addictive or intoxicating drug.

Rational Emotive Behavior Therapy (REBT) and SMART Recovery

I realized I was watching a secular and Humanistic alternative to petitionary prayer at work while sitting in a SMART (Self-Management and Recovery Training) meeting. (SMART is an addiction recovery training method based on scientific research about addiction along with Rational Emotive Behavioral Therapy, designed by twentieth-century psychologist and Humanist Albert Ellis.) SMART teaches its participants strategies for gaining self-control, overcoming perfectionism, and managing anxiety, including my personal favorite kind of anxiety—anxiety *about* anxiety. Rather than offering reliance on a higher power as a way to find inner strength and calm, SMART encourages reviewing REBT "coping statements" like the following:

> I don't have to make myself anxious about anything, or put myself down if I stupidly and foolishly *do* make myself anxious.
>
> I can bear—and bear with—anxiety; it won't kill me.
>
> The world doesn't have to make it easy for me to get a handle on my anxiety.
>
> It is not necessary to be in perfect control of my anxious moments.

To demand that I be in control only multiplies my symptoms.

I don't have to be the one person in the universe to feel comfortable all the time.[11]

That last one gets me every time.

Still, as much as Albert Ellis is widely considered to have been a great psychologist, and his technique has been of benefit to millions, the nature of Humanism is such that we don't simply assume that what works for some will work for all. Ellis's approach involves examining both one's thoughts and one's emotions—thus the "Rational Emotive" in REBT—but some have found the emphasis on reason to be a little too strong, particularly for those moments of extreme emotion when the entire central nervous system seems to be ringing with the terror of a fire alarm that is not a drill. For such moments, what I'd add to the above is that if you need to be crying the whole time that you're saying those things to yourself, it's more than fine. Let yourself experience your feelings! And that goes for men as well as women.

With that in mind, try a little exercise from REBT/SMART that I'd also describe as a wonderfully Humanistic alternative to prayer—one you can use anywhere, anytime, alone or in a crowd, not only to battle craving a drink or a drug but regardless of whether you've ever had any addiction issues. This exercise, called an "ABC," may help you cope with just about any negative emotion or painful situation. The central idea behind it is that our emotions and behaviors are profoundly connected to our thoughts, so by changing your thinking, you may be able to positively affect your future thoughts and actions.

A is an *activating event*, or *adversity*.[12] Unpleasant things happen to us, and provoke a response. We may not be able to control **A**. It could be an earthquake, or just bad weather—try controlling *that*, as SMART trainer Jonathan von Breton has said. It can be something as serious as the death of a loved one, or the breakup of a long-term relationship, or it can be one of those gnawing little things that can bother us beyond all proportion at times, like a flat tire or running late for a meeting. Let's say you're fired from a job. You can't just tell this kind of activating event: "Stop—I'm busy today and don't have time for something lousy to happen to me." Job loss is a legitimate cause for feeling down, scared, or both—but it doesn't mean that drinking to excess or another such negative behavior is a good thing, and it doesn't mean we need to pray for help.

B is the *belief* you have about the **A**. Do you take a rational, realistic view of the situation, or do you imagine something irrational, compounding your pain? For example, if you get fired, you may head down the road of believing it's because you're no good whatsoever. You may begin to believe, regardless of reality, that this sort of thing *always* happens to you, and always will.

C stands for the *consequences* of your beliefs about **A**: both emotional consequences and behavioral ones. In other words, if you get fired and start to believe that losing your job is evidence you're no good, that belief is going to have some concrete effects—some consequences—in your life. You may feel depressed, frustrated, and anxious. And maybe you take a drink you'd be better off not taking, to try to wipe your mind clear of the anxiety. Maybe you scream at your spouse. What consequences can you think of for beliefs you've formulated about negative experiences in your own life?

D stands for *disputing* your own irrational beliefs so you can bring about better consequences. So rather than just allowing your negative thinking to go on and on uninterrupted ("I lost my job because I'm no good; this kind of thing always happens to me . . ."), you make an effort to see the other side. You remind yourself that lots of people get fired at some point in their lives. And maybe this presents an opportunity to get some new training, or try a new career field. And maybe it's okay to be feeling really, really sad right now, and you'll survive that, and it doesn't mean you're always going to feel this way. And you make an effort, with the help of the resources at smartrecovery.org and/or a good therapist, to remember that while you may be feeling lousy today, maybe those bad feelings will pass away a lot faster if you just let yourself have them now, rather than try to push them aside with a drink or some other distraction.

At the very least, it's extremely important to dispute all-or-nothing thinking such as "I'm no good" or "This always happens to me." Maybe you really didn't perform very well on this job. You may have made some serious mistakes, or perhaps you don't have a ton of natural talent in the field you've been in. But it's extremely unlikely that this makes you *all* bad, as we can often catch ourselves thinking if we pay attention to our thoughts. Surely there are other areas in your life where you're getting results that aren't quite as horrible. What are those? This practice of being honest with yourself, not going overboard with either self-praise or self-blame, is a major factor in developing the kind of inner strength that comes from—

E, or *effective new beliefs*. By disputing irrational beliefs, you can take yourself back to step **B**—you can form new, healthier beliefs in response to the challenging events in your life that you may or may not be able to control. Instead of a bad circumstance in your life forcing you to believe bad things about yourself and then feel miserable about those negative beliefs, disputing your own self-destructive thinking can free you to form a healthier, more sustainable understanding of your life. So now maybe you're feeling sad and anxious and angry about getting fired, but you also recognize that nobody is perfect, to err is human, sadness and anxiety and anger are part of life, and they will pass eventually—particularly if you're prepared to start planning for how you can get new training, build on some of your strengths and interests, and make some new connections that can eventually lead to a new job.

Of course, Rational Emotive Behavioral Therapy isn't the only option as an alternative to prayer. What we sometimes need, more than any particular spiritual or psychological self-help technique, is simply another human being we can call on for help with our self-destructive urges and habits. But there are also moments when we don't have the option to call on another person—or a God that can answer. We all need to look inward for help at times. And if we can first understand and accept our emotions and then discipline our reason at such moments, good things will happen.

Meditation and the Relaxation Response

There are a number of other individualized and secular techniques for dealing with anxiety, depressed feelings, and stress. I won't discuss two of the more obvious ones here: namely, drugs such as Prozac and therapy in general, as I'm not a psychiatrist or psychotherapist, and such matters are best left to licensed professionals who can talk with you about your individual situation and see what makes sense for you. I will say, however, that I've been surprised by the extent to which even the kinds of highly articulate and intelligent people I tend to come into contact with often need a lot of encouragement before they'll go find a therapist to check in with, even about a major issue that's come up in their lives—be it a big breakup, a death, an illness, or some unexpected failure at work or school. You can hear about self-esteem all your life and it can still seem more taboo than talking with friends about sex or drugs to pick up the phone and call a referral service for a psychologist; probably because with great talent often comes a feeling that our success defines us, and that it consists of being powerful, in control, and "normal" all the time. But that's not strength, and it's not Humanism either—maybe it's robotism.

May I suggest applying my mother's broken arm test? People break bones all the time. We all know they heal over several weeks or months, that a broken bone can hurt a lot and be a significant inconvenience, but there's no reason to obsess that it will never heal or that it reflects badly on you to have the fracture. Still, never in a million years would you think to say, "Oh, it's no big deal, I'll be fine. It will heal eventually. I don't need a cast. What's the big deal about a doctor, anyway. After all, I have another arm!" If you think you might have the psychological equivalent of a broken arm—something's happened in your life that hurts intensely and is getting in the way of your everyday functioning, though it's definitely not life-threatening and might go away or heal if you just wait several weeks or months—that is not the time to just pray for help, and it doesn't need to be the time to limit yourself to self-help either. Pick up the phone or get online and find some professional help. Again, if the term *professional help* sounds too intimidating, just think of how much of an idiot you'd consider a friend who was too "brave" to go to a "professional" doctor for a fractured wrist.

The next most obvious thing most people would associate with the idea of secular alternatives to prayer is meditation. Of course, you don't need me to inform you that something called meditation exists, and you don't need me to recommend it. You can get that anywhere these days. In fact, following certain areas of our contemporary popular culture and media, you might be forgiven for thinking that meditation was supposed to be a miracle cure of some sort, sweeping away everything from cancer and heart disease to general spiritual ennui.

Meditation can be a part of Humanism as well, though I recommend paying some attention to what kind of meditation you're interested in practicing. If you're not interested in belief in God or prayer, it might be a good idea for consistency's sake to seek out one of the many secular forms of meditation—something involving simply following your breath in and out, rather than a technique focused around reciting the name of a Buddhist or other deity, for example—especially as there's no good scientific evidence showing any difference in effectiveness between the two types of options. But even more importantly, ask yourself: what am I trying to get out of meditating? Many people think meditation is meant to be a way either to *transcend* (Transcendental Meditation is the name used by one of the meditation techniques most aggressively marketed in the Western world) or to be happier. As we discussed in chapter 3, "transcending" the world is okay if it means occasionally detaching yourself from day-to-day worries and concerns, but ultimately, engagement with the world is necessary too—that's where we get the drive that fuels love, strong friendship, and the passion to work for social justice, among other good things. As for happiness—if meditation could reliably produce that, the economy would long ago have come to a crashing halt as people stopped everything to meditate all day. There are no magic formulas, religious or secular, for total happiness. But from a scientific perspective, one positive effect a number of forms of meditation have been demonstrated to produce is called the "relaxation response," which is a little different from what you might commonly think of as simply "relaxing" or chilling out. It's meant to counteract our natural fight-or-flight response.

The Relaxation Response is the name of a best-selling book first published in 1975 by Harvard Medical School professor Herbert Benson. Benson and his colleagues performed some of the first credible scientific research on experienced practitioners of meditation, and began to notice a common result. When confronted with problems we don't immediately know how to solve or respond to, a specific and very old biological response kicks in. It's called the "fight-or-flight" mechanism because prehuman ancestors, when they encountered a potential predator or other such danger, would quite naturally either fight or flee. This instinct worked well for early humans, who also benefited from an adrenaline rush encouraging them to respond quickly with either aggression or defensiveness to problems like rival clans, lions, or lightning storms. Of course, neither fighting nor fleeing is a good option when facing modern stressors such as a meeting with a tax auditor—but though our society has advanced considerably, on this level we are mainly stuck with the brain chemistry of thousands of years ago. And so our fight-or-flight instinct can be triggered many times a day, with no opportunity for us to discharge the energy (in the form of higher blood pressure, faster breathing, increased blood flow to muscles, heart rate, etc.) our bodies produce in response. Over time, our normal level of blood pressure, heart rate, and other bodily functions can increase unhealthily as a result.

But each of the world's religions, along with a number of secular institutions, have over the course of history designed methods for dealing with this—by invoking another bodily response that is, physiologically speaking, the opposite of the fight-or-flight response. Benson identified the commonalities between these various religious and secular techniques—everything from prayers in the Abrahamic religions to Buddhist and Hindu meditation, yoga poses, chanting of various forms, African shamanic trances, and even the "wise passiveness," or "happy stillness of the mind" that Wordsworth wrote about attaining in nature and that we see described by other poets like Tennyson and Emily Brontë.[13] Benson called the physical condition these techniques promote "the relaxation response." In every case we see people taking advantage of the fact that we cannot physiologically be both tense and relaxed at the same time.

If you're a Humanist, atheist, or agnostic who has been curious about meditation or if you've been experiencing negative effects from high stress of late, go ahead and give the relaxation response a try once a day for a week or so, and if you find it helpful, continue as needed. Here are the steps by which Benson instructs you to go about it:

1. Pick a focus word, short phrase, or prayer that is firmly rooted in your belief system.
2. Sit quietly in a comfortable position.
3. Close your eyes.
4. Relax your muscles, progressing from your feet to your calves, thighs, abdomen, shoulders, head, and neck.

5. Breathe slowly and naturally, and as you do, say your focus word, sound, phrase, or prayer silently to yourself as you exhale.
6. Assume a passive attitude. Don't worry about how well you're doing. When other thoughts come to mind, simply say to yourself "Oh well," and gently return to your repetition.
7. Continue for ten to twenty minutes.
8. Do not stand immediately. Continue sitting quietly for a minute or so, allowing other thoughts to return. Then open your eyes and sit for another minute before rising.
9. Practice the technique once or twice daily. Good times to do so are before breakfast and before dinner.[14]

One word of caution: over the years since his initial publication of *The Relaxation Response*, Benson seems to have taken a further interest in religious faith and has published a number of works suggesting that some belief in God or the supernatural may be helpful in eliciting the full benefits of the relaxation response, but the scientific evaluation of such suggestions has been mixed at best. Meanwhile, the response itself has continually been shown to bring positive results when tested. Though some research has shown strong similarities between the benefits caused by the relaxation response and those gained from the placebo effect, this misses the point. The placebo effect is by definition a physical benefit gained passively and unintentionally from a belief that you are receiving helpful treatment. The relaxation response, on the other hand, is precisely about putting your ability to affect your own health and mental well-being to work for you, actively and intentionally. It can be a beautiful example of Humanism in action—sitting down quietly each day, alone or in a group, and focusing on what is good for you, without God.[15]

Art, Nature, and Being Alive Twice

Another important and Humanistic alternative to prayer you don't need me to tell you about—but which is important to mention—is the appreciation of nature and the arts. Just as frequent reminders of the importance of compassion and the golden rule can be helpful (see chapter 4), we secular people can't be reminded too often that art and the natural world are always there waiting for us to appreciate and take part in them. A psychologist friend of mine likes to say that every Sunday he attends the "Church of the Blue Dome." Or as the eighth-century Chinese poet Li Po said to his friend and colleague Tu Fu, "Thank you for letting me read your new poems. It was like being alive twice."[16] What, after all, is making or appreciating art if not taking what we find in the world around us—its radiant natural glory and toxic ugliness, our own love and hate, passion and ambivalence, anger and humor—and transforming it all into something that makes life more beautiful, more worthwhile? One finds this kind of sentiment again and again among great artists and Humanist lovers of art. Katha Pollitt, whom the right wing has labeled the "Atheist in Chief" at *Nation* magazine, has in fact written sensitively that atheism alone, as the rejection of gods and the supernatural, cannot meet our deepest human needs for connection and inspiration, but "perhaps art can go where atheism cannot." And musicologist Daniel Levitin gets at a similar idea in a beautiful chapter entitled "Comfort" in his book *The World in Six Songs: How the Musical Brain Created Human Nature*. The chapter is subtitled with some words a Joni Mitchell fan blurted out to her in gratitude while Mitchell and Levitin were eating dinner together one night. Explaining that Mitchell had helped her get through a rough decade in the 1970s, the fan said, "Before there was Prozac, there was you."

Levitin's story reminds me of a conversation I had with a fan after I quit the rock band I'd been singing with for a few years and announced I was headed to graduate school to study Humanism and religion. "*Religion?*" He asked incredulously, his disappointment in me palpable. "But music *is* religion!" I don't think I mustered much of a response at the time—hell hath no fury like a music fan scorned—but upon reflection, I can say I love music as much as ever today, but the problem with the idea of music as a secular religion is that a concert is not a community. As "cultish" as the fans of some contemporary musical acts or artists can be, such cults rarely get to the point where their members are inspired by their common fandom to support each other in living well and meaningfully, or to come together to help others and make the world a better place.

As the scholar of religion Mircea Eliade explained, modern secular people do have sacred moments and rituals, but they are almost entirely private: "a man's birthplace, or the scenes of his first love, or certain places in the first foreign city he visited in his youth . . . they are the 'holy places' of his private universe, as if it were in such spots that he had received the revelation of a reality other than that in which he participates through his ordinary life."[17] Jonathan Haidt's reaction to Eliade's diagnosis says a lot: "When I read this, I gasped. Eliade had perfectly pegged my feeble spirituality, limited as it is to places, books, people, and events that have given me moments of uplift and enlightenment. Even atheists have intimations of sacredness, particularly when in love or in nature. We just don't infer that God caused those feelings."[18]

Love, nature, and art are all incredibly important; the memorable moments they provide us with are priceless. When I first got involved in organized Humanism I was shocked that the groups I spent time visiting seemed to spend so much time and energy sponsoring debates about the existence of God or publishing magazines, journals, and newsletters, rather than staging poetry readings and concerts or going on hikes together. I hope the coming years will bring a change in that pattern, because we can continue to experience moments of art and nature alone, or in our small and often fairly insular little pairings or groups, but when we do so at the expense of recognizing special moments in other people's lives, and especially at the expense of sharing special moments with others in a broader community, we limit our own potential.

Here are a few ways Humanists and the nonreligious can experience some of the "natural alchemy" that is shared ritual—without actual belief or talk about strange and magical concepts like, say, alchemy.

The Life Cycle

The vast majority of Humanistic and nonreligious people have no idea that formal options exist for secular versions of what we call "life cycle" ceremonies—celebrations marking the major transitions in life, such as birth (think alternatives to baptism, christening, or bris), coming of age (confirmation, bar or bat mitzvah), marriage, and death (weddings and funerals, obviously). But beyond this simple lack of knowledge, there are some who suggest that crafting and holding such events sounds too much like religion. To me, the lack of creativity and originality in such an attitude is shocking.

It's ridiculous to think that giving up belief in unseen beings and the like means we must also give up celebrating important moments in our life in a way that honors family and the best of our traditions and history, seeking to galvanize around our best values and build a good community together. Weddings, funerals, birth, and other such ceremonies not only predate all the currently extant major world religions, they almost certainly predate organized religion itself. And today, there are several organizations in North America, Europe, Australia, and elsewhere that exist to help people celebrate their lives Humanistically.

Baby Naming Ceremonies

"I always thought having a baby was a very private thing to do—boy was I wrong," one mom told on the morning I performed a Humanistic baby naming ceremony for her new son. "I never realized how much help you need, and how much you end up wanting to share the experience with others."

Starting a family is an intensely public thing to do. And the people who care about you want to help—they often just don't know how. A birth ceremony is a chance to let them see how: by loving you and spending time with you. A ceremony can also provide the first major chance since your wedding or commitment ceremony to bring the two sides of your family together and really think about how to blend them most successfully. And of course, it's a chance for you as parents to consider what you want to give to and get out of the process of parenting.

In his Humanist naming ceremonies, Sherwin Wine often said: "When a child is born, hope is born. A child makes us look forward with anticipation and excitement. A child is our link between the past and the future. Children allow us to reach out to the future, with all its possibilities and opportunities. Children unleash the power of our love and creativity and let us discover how much we need to nurture. When we give the gift of life to others, we give life to ourselves. We become alive with hope."[19]

Along with the hope it brings, parenting can also be a daunting prospect, and the fears that often go unspoken are: I don't have enough strength to do it. I won't be good at it. I don't know what to teach my kids. How can I balance being a parent with work and everything else in my life? We acknowledge these fears head-on in our ceremonies. Wine continues, "To love a child is to love life. To nurture a child is to express hope. Children do not exhaust our strength. They allow us to go beyond ourselves and to discover the power of our own creative talents. To be a mother or a father is more than a profession. It is more than a social calling. It is the fulfillment of one of our deepest needs—our need to touch the future and make it live."

Some families choose to do a "welcoming ceremony" where the name isn't a big part of the occasion, but I love naming ceremonies. Why? First of all, a name is significant—choosing the name that a human being will carry for the rest of her or his life is a pretty awesome responsibility. If you don't believe me, if it sounds trivial compared to the stress of submitting your tax forms or getting your next big work project done on time, consider the fact that, as David Brooks has pointed out, "people named Dennis or Denise are disproportionately likely to become dentists. People named Lawrence or Laurie are disproportionately likely to become lawyers. And people named Louis are disproportionately likely to move to St. Louis!"[20]

Seriously, though, choosing a name is often the first major decision a young couple makes together as parents. Every time you do it, you have to negotiate your two different cultural backgrounds, personal styles, family and personal histories (a mom can't give a daughter the name of a beloved great aunt if the dad had a really awful ex-girlfriend with that name). It's a symbolic mutual project involving three people, and it's wonderfully worth those people celebrating together.

You may also want a secular alternative to the moment of choosing "godparents"—beloved and loving friends or family members who pledge to take a special mentoring role in the lives of your children and, movingly, to step in and play an even bigger role should any unexpected tragedy befall you as parents. Why should nonreligious couples be denied an opportunity to formally, publicly acknowledge this sort of commitment? We do often call them "guideparents," acknowledging the guiding role they commit to play in the life of a newborn child.

Marriage

I suspect that a lot of secular and nonreligious people may occasionally wonder: Why even bother getting married in this day and age? Isn't it just some old patriarchal institution, ready for the dustheap of sexism and oppression? Isn't monogamy impossible, and isn't the desire to celebrate it the height of self-absorption?

If you're not feeling ready to get married, far be it from me to pressure you. At the time of this writing, I'm not even married myself, yet. But I've performed quite a few wedding ceremonies, and my colleagues—other Humanist chaplains, rabbis, and celebrants—have performed many thousands of them. It's a fact that there are millions of nonreligious people out there (gay and straight) who are in love and would like to have a meaningful wedding ceremony that suits their values, whether or not they fully understand *why* they want such a thing.

In the past, some wedding ceremonies centered on a groom formally purchasing his bride from her family. Others were primarily intended to seek divine sanction and blessing on the marriage—to ask God to grant some happiness, longevity, and healthy offspring to a couple, often one forced together by the harsh realities of a life in which survival alone was bitter at best and impossible at worst. This is not the place for an anthropological or historical exploration of the history of marriage, though Stephanie Coontz has written a fascinating one for those interested.

Today, though, most people get married by choice. Few of us choose mates primarily on the basis of who can put food on the table or cook it to our specifications. As we often say in our Humanist ceremonies, "the essence of the marriage commitment is the taking of the other person in his or her entirety as a lover, a companion, and a friend. It is a decision undertaken with great consideration and respect by both partners." We want love and companionship. We expect sexual intimacy and attraction, but also a person we respect enough to envision ourselves growing old with. It's a tall order. The wedding is an opportunity to publicly acknowledge that we are blending hope with a grasp of reality—namely, the knowledge that this will not be easy and that we won't be able to have every little thing we might want.

Finally, a Humanist wedding is an important moment in the lives of a couple and their families and friends because it is an opportunity to publicly state that we need help from our loved ones to build a happy home, and to ask for that help. In return, we grant those in attendance a chance to reflect on the role love has come to play in their own lives. Perhaps older couples may be reminded of the hope and intimacy that once energized their own vows; younger singles may gain a model for what they want to find.

As Humanist ceremonies often continue, "Two people in love do not live in isolation from the wider embraces of humanity. To achieve love is not to be absolved of human responsibility. So it is that the institution of marriage is ordained as a public recognition of the private experience of love and as a sanctifying of both parties to its greatest purposes." After the vows are exchanged, I think it's good to ask those in attendance to answer a question: do you, this couple's family and friends, promise to encourage and support them in creating a strong and vital marriage? The answer is always "We do."

Funerals

As much as I try to give religion and religious people the benefit of the doubt when possible, it is almost always a bad idea for a nonreligious family to stage a religious funeral for a nonreligious loved one who has died. At those moments when we are feeling most vulnerable, the last thing we need is false comfort. "I'll pray for you," even nominally believing people suddenly begin to a dying friend. "He's in heaven now," we are told by people who haven't been to church in a decade, about a husband, brother, or son who has just died in pain. If you are among those who do not believe such words true or effective, they can add insult to the pain, because these are not theological statements. They are the words of people who cannot bear to participate in your pain.

For a Humanist, the mourning process begins with accepting that death is real and final and that, with apologies to Epicurus, we fear it. Our fear of death is not only normal, and not to be dismissed, it is part of the motivation we feel to live a good life now, while we still have time. But when the moment comes for someone we love to die, there are almost never good answers for our questions: Why? Why now? To what end? The raw feeling of these unanswered questions is so strong because they are the sign that we care about life, and without our ability to care, sometimes to the point of great pain, we could perhaps continue to walk and speak and chew but not really *live.* And so we do not try to find words that will wish these feelings away—we tell no stories of worlds or reunions or rewards to come—because the mere suggestion that our feelings can be magically washed away is trivializing. And we do not speak of God's presence in the mystery. We acknowledge that no God, no one, no thing can take the pain away except for time, and never entirely.

But there is something else we do—the most important thing. We offer our own presence. A funeral is about people who cared about a common loved one, and who care about themselves and each other, coming together to be present with one another despite the tension and the ambivalence. It is a time to recall the significance of the life that has ended, no matter what it may be—to share stories and memories, meaningful readings and songs, and to express love in the form of laughter and tears, hugs and just sitting. It's amazing that just sitting in the room with a grieving person, neither running away nor wishing her or his pain away, is the single best thing we can do. This is why secular Israelis, with their tradition of building a secular nation from the ground of traditional Judaism up, do not say, "I'm so sorry," at a funeral or, "My condolences," or even, "My sympathies." They say, *"Ani mishtatef b'tzarcha"*—I take part in your grief. As Sherwin Wine put it,

> Death needs courage. It is so overwhelmingly final that it fills our lives with dread and anxious fear. When it arrives at the end of a long and happy life it is never welcome, yet not deeply resented. But when it comes too soon, invading young lives, disrupting hopes and dreams, it adds anger to our fear. We cry out at the injustice of destiny and wait for answers that never seem to come. Courage is the power to confront a world that is not always fair. It is the refusal to beg for what will never be given. It is the willingness to accept what cannot be changed. Courage is loving life, even in the face of death. It is sharing our strength with others even when we feel weak. It is embracing our family and friends even when we fear to lose them. It is opening ourselves to love, even for the last time. Courage is self-esteem. It prefers quiet determination to whining. It prefers doing to waiting. It affirms that exits, like entrances, have their own dignity.[21]

For those readers thinking about a funeral for someone they love, I offer you my encouragement in considering a ceremony that is consistent with your values and those of the person you are grieving. It can be difficult: there may be worries about loyalty to extended family or to tradition; you may worry about propriety and what neighbors will think; or most likely, you simply may not know how to go about arranging anything but a traditional observance. It does help for many reasons to have a trained professional such as a member of the clergy on hand. Though this is not a how-to book, and I can't take you through a detailed set of instructions here, several resources are listed at the back of this book, including whom you might call and what other books and materials you might read. For now, though, if you want to do it independently, you can. I suggest recruiting a strong, responsible, and sympathetic family member or close friend who is not part of the immediately grieving family to help arrange the following and coordinate with a funeral home or memorial chapel:

1. Gather family and friends together before the ceremony to share stories and memories for an hour or so. Laugh and cry. Have your helper take notes or record the conversation.
2. For the ceremony, gather some readings together—poetry, songs, or prose that is relevant to your loved one.
3. Ask one or more people to share stories about the deceased at the ceremony, including telling his or her life story; but there should be no pressure for individuals to speak if they are not so inclined. If the death has been shocking or tragic, acknowledge it. There are special readings for such occasions. You're not alone in going through this.
4. Allow time for silent reflection, and make it clear that those who wish to pray privately at this time should feel comfortable doing so.
5. You might choose words that echo traditional language for key moments of the service. Here is one adapted from Jane Wynne Wilson's suggested committal reading in her excellent guidebook *Funerals Without God:* "We now come to the final moment

in the physical existence of Donald Burke, with respect, honour, affection, regard and love. His passion and intelligence we commit to our memories. His humanity and caring we commit to our hearts. His body we commit to be burned and returned to the cycles of nature he understood so well. 'Earth to earth, dust to dust, ashes to ashes.'"[22]

Holidays

The large golden and mahogany crucifix shone brightly in the stage lights. The church pews were filled to bursting with well over a thousand eager attendees. And during his acceptance speech for the Lifetime Achievement Award in Cultural Humanism we were presenting him with, Sir Salman Rushdie made the following very intriguing comments at our 2007 conference, "The New Humanism":

> I grew up in a community in Bombay, in our immediate neighborhood were people of every conceivable religion and no religion. There were children of Christian, Hindu, Muslim, Buddhist, Sikh, Parsi, and a few like my family, unbelieving households, and we all simply agreed to—it somehow canceled, they all canceled each other out, except that we all decided that what we would do is have each other's holidays as well as our own.
>
> So what it meant to us was more holidays, and this led me, in later life, to wonder where was the holiday for people who didn't believe in God? Schools—the schools my children went to—would say with great pride, in a spirit of ecumenicism, would observe the holidays of all these different religions, but I would say, "But where's the one for the unbelievers? Where is the Kwanzaa for the atheists?" Surely we could make one of those up—Atheistmas. We were—Steven [Pinker] and I were inventing this backstage. So I propose that. Maybe this is it. Maybe that's what this weekend is: it's Atheistmas.
>
> Yes, Amen or similar.[23]

Said as only Rushdie could say it. At the next morning's plenary session, however, Sherwin Wine sat on a panel with Rushdie and raised an objection meant only partially as a joke—that Atheistmas was a great idea, but it wouldn't quite work. Why not? "Because atheist Christians want to celebrate Christmas," Wine said, "and atheist Jews want to celebrate Hanukkah."

We'll leave the angrily baffled questions about how there could be any such thing as an atheist Christian, one of which I always get when bringing up this anecdote, for the next section, on culture. For now we'll simply address the issue that celebrating holidays is a natural, welcome, necessary part of human life, and a Humanism or atheism worth its salt does not callously or humorlessly dismiss this need. In fact, atheists and agnostics *are* interested, given the opportunity, in celebrating the holidays of their childhood, and there's no reason not to, but it can be a refreshing relief to find ways to do so that don't feel so inconsistent with one's beliefs. We'll also briefly address the gist of Rushdie's questions: are there some unique times and ways that Humanists and secularists celebrate their own beliefs throughout the year?

Festivals of Light

Christmas and Hanukkah are really all about sympathetic magic. Our ancestors watched the world get darker and colder every day at that time of year, and every year the cycle provoked their fear. Would there be enough to eat? Were their clothes and homes warm and sturdy enough to survive? Would unseen enemies take shelter in the dark, lying in wait? Besides, seasonal affective disorder (SAD) is real—many of us are more likely to experience depression when there is less sunlight. And so they learned to play tricks on the gods. Thus were born traditions like lights on the Christmas tree and other celebrations around the winter solstice. The Hanukkah tradition of one additional light each day was the ultimate trick—subtly instructing God about exactly what he was supposed to do after the solstice. More light each day, not less. Historians and anthropologists can tell us that the mythological miracle of Temple oil burning for eight days instead of one was clipped onto the observance of Hanukkah long after the holiday had come into existence.

And so, why not take some time out when it is cold and dark for a celebration of light? The traditional story of Jesus can be told as a myth, like any in a book of ancient Greek tales of gods and mortals. Presents can be exchanged, including the heartening recent tradition of choosing charitable gifts in honor of family and friends. The atheist leader Margaret Downey has, in recent years, created a media sensation in her native Philadelphia and offered an inspiring suggestion to others around the country by arranging for a "Tree of Knowledge," elaborately decorated with favorite book covers (including books about atheism and Humanism), to be placed on Philadelphia city government property around the holidays, alongside their displays of Christian and Jewish symbols. And in 2001 Humanist activists Joe Fox and Gary Brill founded a new tradition—HumanLight, a holiday now celebrated by secular families across the nation. Even Santa can come along for the ride too—though consider mentioning to the kids that once upon a time he used to dress in many different colors, until Coca-Cola's massive advertising budget helpfully clarified that he comes down the chimney only in their colors. That's right, the original war on Christmas was not fought by atheists but by American Protestants in the early colonial days who objected that only Papists would indulge in such a thing—and the various commercial accoutrements were gradually added on to make the holiday more palatable to them.[24] Hanukkah can be a chance to celebrate the season, or Jewish culture, or both—the Society for Humanistic Judaism provides resources on how to do such things. In any case, the holiday has only come into special prominence in modern times in response to the unavoidable commercial explosion of its Christian counterpart.

Festivals of Life and Renewal

Then the earth comes back to life. Nature resurrects itself. That is the real story of Easter, and it's why the Easter and Passover season is probably my favorite of the year if I had to choose. Every year I'm amazed by the way it happens again: no matter what is going wrong in my life, the season almost magically provides reasons to view life sunnily. That's just how nature works. It finally provides more light, more warmth, more life, every year. So why wouldn't we want to tell old stories that capture the power of the experience with their metaphors of a man rising from the dead, a people escaping from slavery? I love the family gatherings around these holidays every year, which in fact can also feel as if my family is being resurrected, because I usually haven't seen most of them since December or longer. Without the excuse of the holiday, we would probably not feel enough pressure to actually all show up on a nearly annual basis. And so I love gathering for Humanist Passover Seders, where we read from the Biblical book of Exodus—as literature—and pick it apart over dinner, talking and debating about which values in it we should accept today and which ones we reject. (Freedom from slavery is great, for example; praying that God will visit a series of horrible plagues including infanticide upon one's enemies—less so.) In many ways, Humanists and the nonreligious can celebrate holidays like these by adopting some of the customs that our most liberal Jewish and Christian neighbors have pioneered, then taking things a step further—removing the prayers that might have been said thoughtlessly and without intention anyhow, and substituting little rituals to highlight the modern significance of the occasion, like the new Passover tradition of dipping a finger in one's wine ten times and spilling drops of "blood" for ten modern-day plagues like homelessness, child-trafficking, and nuclear proliferation.

This pattern holds true for many of the other holidays that help us mark and appreciate our year. Humanists can look beyond the mythological trappings and crass greeting-card commercialism and think about why people would have needed to invent such a holiday in the first place—then celebrate *that*. The Christian Lent, Muslim Ramadan, Jewish Rosh Hashanah and Yom Kippur, and the now secular New Year's Eve with its resolutions are times for moral reevaluation. How have we been doing over the past year? Opportunities for atonement—not to God, but to our fellow human beings—are to be taken seriously, for the mistakes we are forever making and, we hope, learning from. And finally, there are a few holidays we Humanists ourselves have created, the most significant of which is probably Darwin Day, which is now honored in thousands of places around the world each year on February 12—Darwin's birthday—as a global celebration of science and humanity. Many progressive churches, recognizing the importance of standing up for belief in reason and evolution against their fundamentalist brethren, join in. HumanLight, an explicitly Humanist alternative to the solstice holidays, incorporates some similar traditions such as gift giving and creative banquets. And if all these aren't enough, create your own rituals, and share them with your community. The single most salient point is that human life, for all its imperfections and disappointments, is well worth celebrating.

Culture

Anthropologist Clifford Geertz wrote, "we are . . . incomplete or unfinished animals who complete or finish ourselves through culture—and not through culture in general but through highly particular forms of it: Dobuan and Japanese, Hopi and Italian, upperclass, academic and commercial."[25]

Ultimately, culture can be many things—the attitudes, practices, values, customs, diet, music, and aesthetic sensibility associated with a particular group of people, whether that group is ethnic, national, or organizational. But perhaps it is best understood, for our purposes, as the opposite of being a "Citizen of the World." Atheists and Humanists have leaped to embrace the laudable notion that all people are equal—equally deserving of treatment with dignity, of human rights. But does universal equality have to mean universal sameness? Could it possibly, even if we wanted it to? In *The Ethics of Identity,* Princeton philosopher Kwame Anthony Appiah poignantly notes that world citizenship becomes problematic when it almost inevitably conflicts with each person's particular identity. Appiah argues that while a universal identity may be a worthy ideal, we must acknowledge that each time we enter a new culture or country, we cannot help but bring with us the baggage that is our roots. He suggests that a worthy alternative goal would be a kind of "rooted cosmopolitanism."

Humanism and atheism often lose out to religion not because of anything remotely related to theological belief, or even because people need to think of themselves as *better* than others, but because we often cannot help but think of ourselves as part of a valued particular group, which is often associated with religion. Whenever this starts sounding too theoretical to me, I think back to an old friend of mixed white Protestant background who, when I was wavering about whether I wanted to become a Humanist rabbi because I wasn't sure how big a deal I wanted to make out of my own

Jewish heritage, said enviously, "You're so lucky to have a *culture*—I'm *nothing*!"

Being good without God shouldn't have to be about becoming *nothing*. We too can have Appiah's sense of rooted cosmopolitanism.

Atheist Anglicans, Cultural Catholics, and Other True Christians

As I am Humanist by faith, a Jew by cultural heritage, and a Humanist chaplain and rabbi by profession, you might think I am commenting on Christianity as something of an outsider.

Not necessarily so. I am also an American, and while I repudiate the notion that this country was founded on or is beholden to Christian principles, I have been influenced—usually for good—almost everywhere I've been across the country, by America's profoundly Christian *culture*. Like the billions of people around the world who watch our elections with profound interest because their fate will be affected by the outcome regardless of where they live, I have always watched American Christian culture with the deep curiosity of one whose life will be shaped by its twists and turns.

Of course, the obvious questions will be: Is Christianity a culture or just a belief? If one ceases to believe in Christian dogma, does one not cease to be a Christian? Never mind that even Bertrand Russell, in his famous essay "Why I Am Not a Christian," allowed that he *was* a Christian in a demographic or cultural sense.[26] I had a fascinating conversation on this topic last year at the Shrove Tuesday Pancake Supper at Harvard's Memorial Church. After eating my fill of delicious blueberry pancakes, traditionally offered as a treat before the austerity of Lent, I joined the church choir director and several students in standing around a piano and singing "Que Sera Sera," and then talked to my friend the Reverend Jon Page, a charismatic, young, liberal Protestant minister, about the cultural differences among Christians. To Jon, a religious history major at Harvard and a graduate of Yale Divinity School, it's obvious that Christianity is as often a culture as a religion. You can see it not only in the kinds of traditions represented by our dinner itself, but also in the clear sociological differences that separate American Protestant groups much more than does theology: Baptists tend to be poorer,

Methodists more middle class, Presbyterians wealthier; Congregationalists tend to trace their roots directly back to English Puritans; one could go on.

No less a personage than Richard Dawkins has called himself a "cultural Christian" of late, along with many other prominent atheists such as the writer Sarah Vowell, who notes that her Pentecostal upbringing affected her deeply, as did her reading of the Bible (in her church, she writes, "we actually read the book") and talks about Martin Luther King Day as a holiday for secular Christians.[27]

Unitarian Universalism

The "liberal religion" of Unitarian Universalism (UU) at times functions as a kind of Humanist Christianity in practice—many UUs are Humanists and many Humanists are involved in UU. Unitarian Universalism formed in 1961 through a merger of two very liberal Christian denominations, the Unitarians and the Universalists. Some of those who orchestrated the merger, particularly among the Unitarians, were hoping the new Unitarian Universalist Association (UUA) would become the nation's first movement of Humanist congregations. That isn't quite how things worked out. Unitarian Universalism has described itself since its inception as a religion without a creed, and thus open to both theists and Humanists of a number of varieties—as long as they self-identify with Unitarian Universalism and its liberal tenets, such as gender egalitarianism, nonracism, and nonethnocentrism. Thus if you go to any of the eleven hundred–plus UU congregations across the United States, what you will find there will vary tremendously from community to community. Some will simply be liberal churches. Others will be Humanist meetinghouses. In others you'll find a mix of practices with a heavy emphasis on Buddhism or paganism.

Former UUA president and historian of religious Humanism William F. Schulz has written of his "guilt" that Unitarian Universalism is one reason Humanists have not had more success building their own unified movement. Schulz argues that the original religious Humanists, despite their allegiances to Unitarian and Universalist groups, intended to build their own independently organized movement of congregations, but ultimately came to rely on the UUA's congregations, where they have lost their voice among other more theological groups. Even if nearly half of the two hundred thousand Unitarian Universalists are self-professed Humanists (the American Humanist Association has generally been smaller), Schulz admits that those Humanists no longer feel as comfortable in a movement that in recent years has been dominated by a more theistic leadership and general character. When I recently attended one of the UUA's impressive national gatherings, I was struck by the two opposing narratives I heard regarding the presence of Humanists and atheists in the movement. On the one hand, I heard a story about a Humanist who stood up at a large conference question-and-answer session and vented his frustration at what he saw as the excessive godliness of much current UU practice. At the end of his remarks he asked, "Are you going to miss us when we're gone?" On the other hand, there was the speaker's response: "Why can't you just let us have our metaphors?"

Black/African American Humanism

One of Unitarian Universalism's most difficult moments came in the late 1960s, when most of the black delegates to the UUA unceremoniously cut their ties with the movement over what they essentially saw as its insufficient progressiveness with regard to black leadership and its desire for separate black UU institutions.[28] This was a damning attack to level against a movement so eager to see itself as racially and otherwise universal, but it was also perhaps one of the strongest proofs that Unitarian Universalism embodies the culture of a particular demographic group—mainstream, liberal, highly educated, white, Judeo-Christian Americans.

African Americans may be largely Christian, but despite having "religion" in common with many whites, they are obviously a cultural group unto themselves. Black American religious institutions exist not merely to promote Jesus's teachings, or Muhammad's for that matter. Rather, they play the crucial role that they do in their community because they provide it with a place of its own to congregate and celebrate itself and its history. The black church has played a unique role in black American cultural life: for decades after slavery, it was the sole black institution that racism did not hinder, and this generated an intense loyalty. As much as African Americans have valued integration, to suggest that they should become so Universalist as to give up the kinds of intracommunity ties that the black church has come to represent would be an affront, not even a very Humanistic thing to ask.

Granted, it is also the case that black Americans have among the highest rates of professed belief in God of any American demographic group. Nevertheless, Anthony Pinn, a prolific writer and professor of religion at Rice University, eloquently champions the concept of African American Humanism. Pinn's writings, such as *By These Hands: A Documentary History of African American Humanism* and *African American Humanist Principles: Living and Thinking Like the Children of Nimrod*, achieve a rich demonstration that Humanistic thought can be found throughout African American history, from the end of slavery—Frederick Douglass, W. E. B. DuBois, Zora Neale Hurston, and many others—through the civil rights movement, whose ideological underpinnings, he argues, were "clarified through attention to humanist principles."[29] Pinn points out that A. Phillip Randolph, a truly great labor leader by any standard, who was known as the "dean of Negro leaders" right up to the time Martin Luther King Jr. stepped up to the national stage, was a committed Humanist, even signing the American Humanist Association's Humanist Manifesto II in 1973. And he illustrates the creative spirit of black Humanism by selecting wonderfully freethinking passages from important African American literary works, such as this excerpt from Richard Wright's canonical memoir, *Black Boy*:

> One boy, who lived across the street, called on me one afternoon and his self-consciousness betrayed him; he spoke so naively and clumsily that I could see the bare bones of his holy plot and hear the creaking of the machinery of Granny's maneuvering.
>
> "Richard, do you know we are all worried about you?" he asked.
>
> "Worried about me? Who's worried about me?" I asked in feigned surprise.
>
> "All of us," he said, his eyes avoiding mine. "Why?" I asked.
>
> "You're not saved," he said sadly.
>
> "I'm all right," I said, laughing.
>
> "Don't laugh, Richard. It's serious," he said.
>
> "But I tell you that I'm all right."
>
> "Say, Richard, I'd like to be a good friend of yours."

"I thought we were friends already," I said.

"I mean true brothers in Christ," he said.

"We know each other," I said in a soft voice tinged with irony.

"But not in Christ," he said.

"Friendship is friendship with me."

"But don't you want to save your soul?"

"I simply can't feel religion," I told him in lieu of telling him that I did not think I had the kind of soul he thought I had.

"Have you really tried to feel God?" he asked.

"No. But I know I can't feel anything like that."

"You simply can't let the question rest there, Richard."

"Why should I let it rest?"

"Don't mock God," he said.

"I'll never feel God, I tell you. It's no use."

"Would you let the fate of your soul hang upon pride and vanity?"

"I don't think I have any pride in matters like this."

"Richard, think of Christ's dying for you, shedding His blood, His precious blood on the cross."

"Other people have shed blood," I ventured.

"But it's not the same. You don't understand."

"I don't think I ever will."

"Oh Richard, brother, you are lost in the darkness of the world. You must let the Church help you."

"I tell you, I'm all right."

"Come into the house and let me pray for you."

"I don't want to hurt your feelings . . ."

"You can't. I'm talking for God."

"I don't want to hurt God's feelings either."[30]

The concept of "redemptive suffering"—that earthly suffering is desirable as an end unto itself because it leads to redemption after death—is, for Pinn, the strongest reason to cut ties with the black church. He paints a painful picture of the harm it has done a black community that has been subjected, and at times subjected itself, to so much intense and *unneces-*

sary suffering. We can only imagine what might have been if more African Americans adopted the Humanist stance that suffering is purposeless and to be struggled against, toward the goal of a more dignified life for all. Still, despite the solid intellectual and theoretical foundation Pinn lays out, he sees himself as a "Humanist in search of a home."[31] He does not have the congregational equivalent of a black church, but wants one: "Regarding this, I agree with Cornel West—institutional affiliation helps ground the intellectual's role in social transformation."[32] Perhaps under an African American president who is also in search of a congregation, and who has written eloquently of his late mother's Humanism, some progress can be made on this front in the coming years.

Other Cultural Humanisms

Because the stream of goodness without God runs through every society and culture, the potential for a "cultural Humanism" exists everywhere we might look. Amartya Sen, in *The Argumentative Indian* (2005) and *Identity and Violence* (2006), has sketched the outlines of an Indian identity thoroughly informed and inspired by Humanism from the Carvaka and Lokayata through modern times. The Indian city of Vijayawada, Andhra Pradesh, has had a very active Atheist Centre since the days of Gandhi, who was an admirer of its founder. The Atheist Centre has since those days sponsored a variety of programs combating social inequality and injustice, providing free education in science and the humanities to children of Untouchables and other poor classes, and otherwise breaking the often oppressive hold that conservative Hindu traditions can have on rural, underprivileged Indians. Gandhi was in fact well aware that he occasionally lost volunteers in the region who would leave his compound to go and work with the less famous Gora, the founder of the Atheist Centre, because Gora was doing more good than Gandhi for the Untouchables.[33] There are also organizations like Indicorps—a cultural and service organization for members of the Indian diaspora, started by social entrepreneur Anand Shah and his two siblings just after the millennium. Shah has said, "In order to transcend my identity, I must first understand it." I heard about this very Humanistic institution from Harvard student Prerna Srivastava, who talked about how it reinforced for her that an Indian and Hindu heritage is not about God or gods but about "community, service, interconnectedness, and relationships."

The Muslim apostate Ibn Warraq (a pen name taken by Islamic infidels over the years) has been prolific and increasingly known in recent years for his books describing the process of turning to Humanism from an Islamic perspective, and of attempting to bring about an "Enlightenment" in contemporary Islam by means of critical scholarship on the Qur'an, the history of intellectual and theological liberalism in Muslim circles, and on those who have "left Islam" as a dogmatic system of supernatural beliefs and accompanying rituals.[34]

At the New Humanism conference, I asked Salman Rushdie to help demonstrate this fact by talking about the roots of Humanism in the Islamic world, and he gave an elegant account of an authentic stream of Muslim heritage that can be traced back to the democratizing tendencies of Cyrus in ancient Persia and through the great medieval philosophers like Averroes, as we mentioned in chapter 2. Rushdie also recalled that this tradition extended into modern times, and sounded a cautiously optimistic note about its potential revitalization:

> I grew up in a world in which it seemed perfectly normal to suggest or to posit a cultural description of Islam rather than a religious one. This wasn't just limited to my aged family [although] one has to admit that it was a view probably most widely held among the highly educated and economically more affluent Muslims. Nevertheless, as an idea it was very much alive—so much so that my father could discuss something that is perfectly obvious when you read the Qur'an, which is that the chapters have got jumbled up; that what is now the orthodox, unarguable-beyond-dispute arrangement of the chapters of the Qur'an is quite evidently wrong . . . [T]he key question, then, is: how can we assist in the reopening of the subject of questioning? I don't think you can simply say that Islam, as it is now, is fine—just find the liberal Muslims, the tolerant Muslims, and talk to them. There are great intellectual problems here in regard to how one can approach the subject of religion, the subject of ideas, and those need to be solved from both inside and outside.[35]

Tu Weiming, a scholar of Chinese history and philosophy and of Confucian studies, has devoted his internationally decorated career to making the case for Confucian Humanism: an affirmative, spiritual vision of Chinese and East Asian identity that makes an explicit commitment to universal human rights, ecological conservation, ecumenical dialogue, and progressive capitalism (or at least progressive anti-Communism).[36] And noted French Buddhist scholar Stephen Batchelor makes a passionate plea for agnosticism and naturalism as not only a historical element of Buddhist philosophy but also an important aspect of contemporary Buddhism, in his recent best-selling work, *Buddhism Without Beliefs*.[37]

Humanistic Judaism

But if the theme here seems to be one of scholars and public intellectuals issuing erudite white papers about lofty ideas that never see the light of day-to-day practice, there is a reason to take heart.

Rabbi Sherwin Wine was a brilliant young reform rabbi in suburban Detroit when, in 1963, he rocked the American Jewish world by founding the first-ever synagogue without God. He'd known he was a Humanist nearly all his life, but loved the Jewish culture and community he grew up around, and loved the job of rabbi as held by the modern, liberal, elegantly dressed and groomed men who served his family in that capacity at their Conservative synagogue. The position of rabbi in affluent suburban Detroit around the time of World War II, after all, hardly revolved around offering devotions to God, studying obscure Talmudic texts, or henpecking Jews about following the laws of kashrut (how to keep kosher). All that crowd expected from its rabbi was a stately presence and some sage wisdom when it came time to get married, hold a funeral, or bless a baby's coming into the world. The rabbi was expected to give an inspiring sermon, along the lines of a presidential speech, on the few times a year when most Jews bothered to pass through the gilded old temple. He was supposed to set an example. And recite a few Hebrew prayers nobody paid much attention to anyhow.

It was a great gig, especially for a gifted showman like Sherwin, who loved people and loved to talk to them. Sherwin decided against a career in academia ("Teaching wouldn't have fit with the intensity with which I work. I wanted my own platform, my own venue")[38] and instead became a Reform rabbi (and then a Jewish chaplain in the army serving in Korea), not in spite of his atheism, but in fact because Reform Judaism was the most liberal form of Jewish community in America back then. Many Reform rabbis were atheists, though they followed a kind of don't-ask, don't-tell policy about it with their congregations.

But Sherwin refused to accept don't-ask, don't-tell-if-you-don't-believe. He wanted to say it out loud. And not just "There is no God." More importantly, he had no stomach for wasting time in front of his congregation mouthing words no one took seriously ("Blessed art thou, oh Lord our God, ruler of the Universe, who sanctified us with his commandments . . .") when he could be talking to them about things that really mattered: What does it mean to live a good life? What do we believe in, if not God? How do we cope with death, tragedy, and the absurd unfairness of life? How can we find the strength to be happy in the face of all the unhappiness around us?

So Sherwin gathered a group of people who were dissatisfied with the congregational options currently at their disposal and told them he intended to do something new and different. He wanted to leave his synagogue behind and start a new one, meant to function as a community center for people, not a house of God. He recognized that Judaism is a culture, and can claim famous doubters from Freud to Theodore Herzl to Woody Allen, not to mention the 49 percent of American Jews today who say they are not Jewish by religion. Wine preached powerfully that we might take pride in our culture while affirming the equality of all human beings as part of a Humanist worldview. In this new synagogue, the philosophy, the message, and the driving force would not be Jewish theology, nor would it be any God at all. The congregation would focus on meeting human needs, especially the need to strive together for human dignity. Eight couples went with him to Birmingham, Michigan, and formed something called the Birmingham Temple. Before long, *Time* magazine was decrying the "atheist rabbi" in Detroit.

When I met Sherwin Wine it was more than thirty-five years later. The Birmingham Temple had moved to nearby Farmington Hills, where it had grown to become a gorgeous, sprawling suburban synagogue for several hundred rebel Jews and their friends. The movement of Humanistic Judaism had taken root and spread to congregations and communities around the world. Sherwin had written several books, such as the classic *Judaism Beyond God*;

founded a Humanistic Jewish rabbinic program to train future Humanist leaders; and publicly debated religious fundamentalists such as Jerry Falwell and Meir Kahane. By that point he'd also performed thousands of weddings, funerals, bar and bat mitzvahs, and baby naming ceremonies based not on praise of God but on celebration of the human spirit. And he'd listened at the bedside of hundreds of sick and dying patients with the same passion, commitment, and warmth that drove him to found several Humanist umbrella organizations nationally and internationally.

He listened to me too. Back then I was still a confused postadolescent, exploring Buddhism and rock music. My father's death haunted me, and my life choices expressed a subconscious whine, "Why me?" Without the theatrics of grabbing me by the collar, he patiently argued that life isn't fair, and the good life is not one of constant sensual pleasure or of narcissistic self-regard. Today I am among those dedicating our lives to carrying on the new Humanistic tradition he helped weave into the older Jewish cultural tradition we are still connected to. As he described it:

> While many institutions in the old authoritarian religion were harmful, not all of them were. Congregations and rabbis were useful inventions. Secular Jews need full-service communities, and they need trained leaders who can respond not only to their Jewish cultural needs but especially to their human needs for coping with the human condition. Suffering and death are also Jewish. Struggling for happiness is also Jewish. In many ways Humanistic congregations function in the lives of their members in the same way as Reform, Conservative, and Reconstructionist synagogues do. They provide the same services, ask the same questions—even though they provide different answers.[39]

Of course, from a Humanistic perspective there is no reason to see even our most beloved heroes as perfect saints. Surely the movement he built is nowhere near as far along in its development today as it could be. There is a lot of work left to be done for those willing to do it. Moreover, he didn't get much of a chance to work on his belief that the combined forces of Humanism and a cultural community could well serve many non-Jewish communities as well. I hope in the coming years we will see more actively organized groups of African American Humanists, Humanist Quakers and Unitarian Universalists, Indian Humanists, Humanist Buddhists, and the like. In any case, we will have a great model for such communities available to us, if we are willing to reach beyond ourselves and acknowledge our individual needs to come together with others and serve a higher human purpose.

Community

It seems that almost everyone who's ever written about society, not to mention sociology, has an opinion on community—what it is, how and why we lost it, who took it from us, how we must get it back—from the Greeks, who defined it as an inclusive state but excluded anyone and everyone who got in the way of their ideal polis; to Augustine, who argued that what we needed was a city of God, not a city of man; to Hegel, who complained that modernity was destroying our community; to Weber, who analyzed *communitas* as belonging; and Durkheim, who diagnosed modern society's need for a new, posttraditional form of that belonging; all the way to the hot windstorm of today's discourse on community—that is, everyone from Rush Limbaugh whining that you can blame it all on the Democrats, to *Bowling Alone,* to deconstructionists and queer theorists who want to blow up yesterday's notions of community and start over again, and everywhere in between. It would be cruelly boring if I went through the history of all this just to make a point that can be put more succinctly. Here it is: Community can be many things at any time for anyone. But whatever it is, we all know we need it. We all must have it.

An Ethical Culture

This book wouldn't be complete without a description of an important historic episode—the story of a man named Felix Adler and the movement of Ethical Culture he founded in 1876.

Adler was born in Germany in 1851 to a Reform rabbi father who immigrated to the United States when Felix was a boy, to become the rabbi of the wealthiest and most prestigious synagogue in New York City. Felix was something of a prodigy and given every opportunity for education, including being sent to Germany for rabbinical training of his own, where he internalized the momentous changes taking place in science and religion, and thus decided that the Reform rabbinate could contain neither his philosophy nor his ambition.

In Germany, Adler decided to build a new religion that would be beyond science without being inconsistent with it. Upon his return home he began talking about it, and stirred enthusiasm in his peers. And in May 1876, not long after his twenty-sixth birthday, Adler took the stage of a rented hall to address his family, friends, and the collection of admirers and curious onlookers that had already begun to gather around him. Adler proposed a new movement meant to address a "great and crying evil in modern society. It is want of purpose."[40] The new movement Adler proposed was intended "to entirely exclude prayer and every form of ritual . . ." His address was gripping, inspiring, and full of Humanistic conviction: "freedom of thought is a sacred right of every individual man. Believe or disbelieve as you list—we shall at all times respect every honest conviction—but be one with us where there is nothing to divide—in action. Diversity in the creed, unanimity in the deed. This is that practical religion from which none dissents. This is that Platform broad enough to receive the worshipper and the infidel. This is that common ground where we may all grasp hands as brothers united in mankind's common cause . . ."[41]

Adler took seriously his call for action and not words. A movement was incorporated almost immediately and began to attract members rapidly. Meetings were set for each Sunday, where Adler would deliver "platform" addresses often outlining ideas for ambitious social service projects meant to correct the glaring injustices of the urban life of that time. He and his followers then got to work assembling those projects.

Within a few years Ethical Culture had created the first free-of-charge Visiting Nurses program in New York's history, to address the frightening tuberculosis epidemic of the 1880s. Young nurses from the Ethical Culture community were sent into deadly city slums by themselves, or with their mothers if they were too young to go alone. The girls were taught that their mission was "to help others not for any future reward, but simply because we thought it was right to do right."[42] They must have made quite the sight for those young men suffering in lonely beds—one of whom told them that he much preferred their visits to those of Catholic nuns "because we brought food to nourish his body while the Sisters were primarily interested in saving his soul."[43] Through the success of Ethical Culture member Lillian Wald, this program was permanently incorporated into the city's network of social services and still does great good today—though, as planned by Adler and others in those days, Ethical Culture asked for and was given neither credit nor further say in the fate of the charity it founded.

A beautiful building was erected on Central Park in 1910 that still stands, the size of a megachurch, as the home of the New York Society for Ethical Culture and its national body, the National American Ethical Union. The building was filled with a superstar cast of congregants and collaborators, including Jane Addams, Lillian Wald, Booker T. Washington, W. E. B. DuBois, William James, Walt Whitman, and Samuel Gompers. Ethical Culture was also involved in the establishment of the National Association for the Advancement of Colored People in 1909, and Adler's successor as head of the movement, John Lovejoy Elliot, was a key player in the founding of the American Civil Liberties Union, or ACLU.

Unmistakably, Adler was a brilliant organizer who knew how to motivate and convince the most talented leaders of his day of the importance and worth of his mission.[44] He also identified strong leaders who then created impressive successes of their own. For example, Stanton Coit, sent to London to develop Ethical Culture overseas, set up an Ethical Culture Church that made more effective use of symbols, ritual, and music than what had thus far been developed in New York. Within several years of his arrival in England, six hundred people were regularly attending his weekly Sunday services.

Another successful project was the free kindergarten Adler and his colleagues founded to offer ethics, science, and soup to the city's poorest children. School organizers, including Adler, would knock on doors in Manhattan's "gas house district" with leaflets and idealism. Parents initially suspected it was some sort of elaborate kidnapping scheme. Unfortunately, the several schools Ethical Culture ultimately founded met the same fate as the Visiting Nurse program—after achieving great success, its founders turned their work over to the general public, taking little or no credit for themselves, for Ethical Culture, or for Humanism.

According to Ethical Culture movement historian Howard Radest, "These early efforts at social service also set a pattern in response to social need. A project for meeting that need would be devised and implemented, and thereafter move toward independence and ultimate lack of identification with Ethical Culture. Perhaps the austerity of Adler's ethic was a disservice to the movement he sought to build, for it guaranteed that much of the wealth, energy, and competence that rallied to Ethical Culture would be put at the disposal of causes outside of the movement itself."[45]

The nomenclature issue reared its head as well: when, later in his life, his movement's next generation began to gravitate toward affiliation with the term *Humanism* and its new organizations, Adler let pride and disagreement over minor differences cloud his cooperative spirit, shunning the term for what he described as its emphasis on "human ends" rather than "transcendental ideals"—even though he surely knew enough about the Humanist movement at that time to know that its ideals were no more nor less "transcendental" than his own.

Sadly, the Ethical Culture community still has not really addressed the nomenclature issue—the divisions between it and Humanism today are so minimal that the fact that some small community Humanist and Ethical Culture groups don't significantly collaborate or even merge, simply because of their different origins and names, is tragicomic. The movement hasn't significantly revitalized its meeting format in 135 years—no religious denomination could ever survive that. And it hasn't changed the awkward name Leader for its clergy. Still, when compared to the beauty and pioneering relevance of Adler's vision for a congregation that could do great good without God, these concerns are minor. They could be fixed relatively easily with the will of current membership and an infusion of new interest.

The Ethical Culture building still stands as a living monument—the largest Humanist congregational gathering house ever built. But more than that, it is a symbol of all the wonderful things Humanists and the nonreligious have done together in the past and can do even more of in the future: articulate scientifically sound and creatively inspiring values; build supportive and loving communities around those values; develop aesthetically powerful rituals and a sense of engagement with culture; serve the community with uncommon bravery and measurable success; and play a leading role in the most urgent social struggles of the age. For all that and more, the New York

Society for Ethical Culture should be a place of pilgrimage, among a handful of must-see American tourist stops for anyone interested in Humanism, atheism, secularism, or world religion generally—along with the Birmingham Temple in Michigan; the Center for Inquiry in Amherst, New York; and the United States Congress (not because Jerry Falwell was right about its being a House of six hundred Humanists, but because any American Humanist can arrange to visit his congressional representative alongside the staff of the Secular Coalition for America, a dynamic group affiliated with organizations such as the American Humanist Association, the American Ethical Union, the Society for Humanistic Judaism, and many more).

Though at the time of this writing, the New York building has some chipping paint, it can still be a huge home for Humanism. What will we do with it? And will we build more such centers in Boston? San Francisco? Atlanta? Austin? Buenos Aires? Barcelona? Beijing? Mumbai? Yes. We Humanists need to build more actual structures of our own. Only a tiny number of atheist and Humanist groups and communities have their own buildings. Can you imagine a U.S. Supreme Court session in some rented elementary school classroom, with the Justices sitting around on those little orange plastic chairs? We will never have a chance to build up the political, charitable, educational, and social work we do if we don't have our own spaces. Finding a Humanist congregation is not some oddball curiosity of an idea; it's not even a luxury, to be addressed after we succeed in getting "In God We Trust" off the dollar bill. If we ever want to be anything more than a downtrodden minority, it is a necessary response to one of our most aching and eternal human needs.

Now I fear I have pushed my allies further than some will want to tolerate. The world of organized Humanism is as diverse as that of organized Christianity or Judaism. It contains many viewpoints on what the nonreligious ought to stand for and do. Many will agree with me that we can be good without God, but strongly disagree with one or another aspect of my vision for a Humanist community. Still, I have the highest respect for and identification with those working in the various loosely connected streams of the Humanist, atheist, Ethical Culture, secularist, and free thought movement today. Their aims, in very large part, are my aims. Their failures are my failures. And I dearly hope that if I am in any way successful as an individual, it will be because I have contributed to their success. I ask their forgiveness and their understanding that it is simply not a Humanist value to suspend one's judgment of history or reality in order to prop up one's friends with hagiography.

For all its flaws and foibles, the entire movement has actually accomplished an incredible amount given that it has existed in an organized sense for, essentially, a single century. Given Humanism's mobilization of ideas and recruiting of intellectual leadership, along with its strong influence on its religious contemporaries worldwide, it is fair to say that no spiritual movement worldwide has accomplished more in its first hundred years. And so it is my aim now to provoke serious thought, not only among religious people as to whether they have given goodness without a God a fair shake, and among nonreligious people as to whether they might do well to affiliate with a movement that represents their ideals and needs their help, but also among active Humanist leaders who are doing what I believe to be heroic work and will not be served by complacency, especially when they stand at the threshold of making a deep and visible difference in the world.

Osama bin Laden's greatest success was in recognizing a generation ago that inequality, disenfranchisement, and anger were where his part of the world was headed, and setting out to organize disaffected young men into a mass movement. As Eboo Patel notes, when bin Laden met Zarqawi and Atta, he saw their entrepreneurial leadership skills first. He saw their potential as lieutenants, their ability to mobilize masses of young people to make his vision of radical Islam a reality. Eventually we will either match this combination of passion and pragmatism, or be overwhelmed by it.

Herding Cats, or "Community Organizing"

For too long, organized secularism has been an oxymoron, like herding cats. People who believed in a more humane, Humanistic world mostly wrote off the possibility of gathering their peers together so that together they could actually have some influence. Weekly meetings were too dogmatic. Trained leaders were too priestly. Dedicated meeting spaces too churchy. Fund-raising too smarmy and reminiscent of the worst evils of the collection plate. Never mind that the very existence of cherished secular institutions like museums, universities, and the like depends on trained leadership, dedicated spaces, and fund-raising.

But it would be one thing if the allergy were to top-down institutions alone. Bottom-up approaches to promoting goodness without God have also been politely ignored. Many local and national secular communities have the same president for decades. They do not have regular elections. They employ no outreach strategy with a prayer of ensuring that many people beyond their insular social circles will ever hear of them.

If all we stand for is on the one hand repudiating top-down religious authority as represented by the clergy, and on the other hand, with noses turned up, condescending toward more democratic, grassroots efforts to organize Humanist communities and rejecting anything that reminds anyone in any way of a church—then that's not just cutting off our nose to spite our face, it's chopping off our head to spite our body.

But wherever there are good people, there are people who believe in being good without God. There are freethinkers wherever there are free people who are able to think. There are Humanists wherever there are human beings.

The Obama campaign swept the United States in 2008 with the help of strategies and techniques that seemed radically new to everyone except those who knew what a "community organizer" actually does. Community organizing is nothing more, nor less, than herding cats. It is offering people the opportunity to come together not out of ignorance, or because they are sheep, but because they are intelligent enough to understand that in a world without supernatural or natural miracles, only we can make the world a better place, and we cannot do it alone. Here is a letter to you, inspired by one of the many messages the Obama campaign sent over the course of its run to the presidency—a run that was unexpected by most, but not too hard to understand for those of us who have studied community organizing and know its power.

> Anyone in the world reading this has the potential to help build Humanism—not just for our own sake or the sake of the individual Humanist communities we will build and strengthen, but as a way of maximizing the impact that each of us as individuals can have on the entire world.

Now, after reading this book, it's time to start sharing your ideas and energy with others that need them. It's time to seek out people from diverse backgrounds and experiences who, like you, are passionately committed to living a good life and building a good society, and who just happen to be among the millions who can soberly face the reality that this work must be done without God.

That's why I am asking you now, whether in conjunction with a local Humanist, atheist, or secular group, or on your own initiative if no such group exists near you, to organize a meeting about Humanism in your home, or elsewhere in your neighborhood.

Around the United States and the entire world, people like you who are committed to goodness without God or traditional religion, whether longtime Humanist leaders or newcomers exploring these ideas, can host house meetings explicitly extending a hand to others who see themselves as Humanists, atheists, agnostics, nonreligious, etc.—whether they've been involved in Humanism before or have never heard of it.

The purpose of these meetings is to meet, get to know one another by sharing the stories of why you've chosen to embrace a human-centered philosophy of life, and then, based on the shared values that emerge from these diverse stories, build a grassroots organization that can spread the word about Humanism around your community. Together, groups like this around the globe can build community connections while fighting climate change; defending church-state separation alongside progressive religious allies; cleaning up parks; tutoring children in need; celebrating holidays, weddings, and funerals; and eventually acquiring their own physical centers that will be able to house community service programs like homeless empowerment and addiction recovery treatment driven by the best practices and latest scientific knowledge of our time; all the while spreading religious freedom by allowing each person to decide for him- or herself whether to embrace a Humanistic or theistic worldview.

Sure, it's an ambitious project and requires some commitment. But it's incredibly exciting to be involved, alongside your friends and neighbors, in a project that has global significance for a cause you care deeply about. And with this book as a beginning, along with Web sites like mine and those of the American Humanist Association and International Humanist and Ethical Union, and the other resources I've outlined in the appendix, you'll have plenty of resources to get you started. We can't do it without you.

Conclusion

I write this as a call to action. The subject is Humanism, but convincing you to become a Humanist or to use that word to describe yourself isn't my goal. If you are not a Humanist, please go in peace. You have my respect. I ask you, for the sake of all humanity, for yours. And if you are a Humanist, and if you've been inspired by this book, please know that that in itself brings me no special joy if "Humanist" to you means merely "one who denies the existence of the gods." Humanists must be known for their actions.

We must act together for our own good and for the greater good. We are so fortunate to have evolved and been nurtured to possess reason, compassion, and creativity. It is what we do with those qualities that will determine everything. The fact that we live without God is, in a sense, not up to us. It's not really a choice. We see the world around us. We use our amazing human ability to think and believe with all our integrity that there is only this one natural world. But goodness is a choice. It is the most important choice we can ever make. And we have to make it again and again, throughout our lives and in every aspect of our lives. We have to be good for ourselves. We have to be good for the people we love. We have to be good for all the people around us, be they friend or foe. We are forced to be good without God. If we can accept that reality and act with courage, we can be very good indeed.

This is the beginning—only the beginning—of a movement that will change our world. I've been told that maybe in the future the world will be ready for Humanism. Maybe someday. But that's not an attitude we can afford to accept. As we can learn from a nearly two-thousand-year-old saying, it is not your responsibility to finish the work, but neither are you free to desist from it.

Let us go out and make a difference now.

Postscript: Humanism and Its Aspirations

The following document is a short statement of common Humanist beliefs and principles. It is called "Humanism and Its Aspirations," and was produced in 2003 by leaders of the American Humanist Association to provide both a starting point for those studying Humanism and an eloquent affirmation of its goals.

"Humanism and Its Aspirations" is also sometimes referred to as the Humanist Manifesto III, because it is a follow-up to a 1933 document called the Humanist Manifesto and a 1973 sequel, Humanist Manifesto II. The original Manifesto was a kind of coming-out party for the earliest group of intellectual and cultural leaders who called themselves Humanists in the sense of being good without God. It was a bold and optimistic—in retrospect probably a bit too optimistic—statement about human potential and responsibility in an age when supernatural religion seemed on the decline. (It was also produced in an era when the word *Manifesto* had many fewer negative connotations.)

The writers of Humanist Manifesto II set out to articulate Humanist principles that could square with the reality of the Holocaust, Communist tyranny, nuclear proliferation, and other horrors of the twentieth century that their predecessors did not predict in 1933. They also wanted to clarify that many Humanists no longer thought of themselves as "religious," as the original Manifesto's authors did. But Manifesto II was a little too long to be an effective summary of Humanist beliefs, and a little too focused on the culture wars of the time and what Humanists opposed, rather than simply an explanation of what Humanism stands for.

This third document draws on some of the foundational language of both its predecessors, but has a simplicity and balance that have made it very helpful for me personally and many in my community.

"Humanism and Its Aspirations" is a statement of beliefs and principles, not canon law or a set of commandments. Humanists just don't work that way. If you're looking for one set of statements that explains everything about Humanism, you're misunderstanding what kind of tradition it is—not to mention misunderstanding human beings. Even if one could stop the world right now and design a system of ethics that would govern every action in ten words or less, that is not what a sane, rational person would want to do. It would be tyranny. Because who would decide what system we were operating on? Humanists do not believe in infallible authorities, and we value the messy, painstaking process of bringing a group of individuals to an evolving, overlapping consensus. In other words, the document below was written by multiple authors and editors over a period of many years (Manifesto II began to be planned about twenty years before its eventual primary author, Paul Kurtz, completed it), and even after all those years of debating and discussing nearly every word, the authors (and readers) still had plenty of disagreements over various aspects of the text.

We celebrate the fact that the Humanism expressed here is not exactly like the Humanism of the past, nor will it remain exactly the same in the future. Humanism is always evolving, and Humanism does not require that all who embrace it agree on everything. Rather, it expects that we will find and acknowledge much common ground that will outweigh our differences. I am grateful to the American Humanist Association and to primary drafter Fred Edwords for allowing their work to be reprinted here, and I invite you to visit the AHA at www.americanhumanist.org to learn more about these and other foundational documents of Humanism.

Humanism and Its Aspirations

Humanist Manifesto III: A Successor to the Humanist Manifesto of 1933*

Humanism is a progressive philosophy of life that, without supernaturalism, affirms our ability and responsibility to lead ethical lives of personal fulfillment that aspire to the greater good of humanity.

The lifestance of Humanism—guided by reason, inspired by compassion, and informed by experience—encourages us to live life well and fully. It evolved through the ages and continues to develop through the efforts of thoughtful people who recognize that values and ideals, however carefully wrought, are subject to change as our knowledge and understandings advance.

This document is part of an ongoing effort to manifest in clear and positive terms the conceptual boundaries of Humanism, not what we must believe but a consensus of what we do believe. It is in this sense that we affirm the following:

KNOWLEDGE OF THE WORLD IS DERIVED BY OBSERVATION, EXPERIMENTATION, AND RATIONAL ANALYSIS.

> Humanists find that science is the best method for determining this knowledge as well as for solving problems and developing beneficial technologies. We also recognize the value of new departures in thought, the arts, and inner experience—each subject to analysis by critical intelligence.

HUMANS ARE AN INTEGRAL PART OF NATURE, THE RESULT OF UNGUIDED EVOLUTIONARY CHANGE.

> Humanists recognize nature as self-existing. We accept our life as all and enough, distinguishing things as they are from things as we might wish or imagine them to be. We welcome the challenges

*Humanist Manifesto is a trademark of the American Humanist Association.

of the future, and are drawn to and undaunted by the yet to be known.

ETHICAL VALUES ARE DERIVED FROM HUMAN NEED AND INTEREST AS TESTED BY EXPERIENCE.

Humanists ground values in human welfare shaped by human circumstances, interests, and concerns and extended to the global ecosystem and beyond. We are committed to treating each person as having inherent worth and dignity, and to making informed choices in a context of freedom consonant with responsibility.

LIFE'S FULFILLMENT EMERGES FROM INDIVIDUAL PARTICIPATION IN THE SERVICE OF HUMANE IDEALS.

We aim for our fullest possible development and animate our lives with a deep sense of purpose, finding wonder and awe in the joys and beauties of human existence, its challenges and tragedies, and even in the inevitability and finality of death. Humanists rely on the rich heritage of human culture and the lifestance of Humanism to provide comfort in times of want and encouragement in times of plenty.

HUMANS ARE SOCIAL BY NATURE AND FIND MEANING IN RELATIONSHIPS.

Humanists long for and strive toward a world of mutual care and concern, free of cruelty and its consequences, where differences are resolved cooperatively without resorting to violence. The joining of individuality with interdependence enriches our lives, encourages us to enrich the lives of others, and inspires hope of attaining peace, justice, and opportunity for all.

WORKING TO BENEFIT SOCIETY MAXIMIZES INDIVIDUAL HAPPINESS.

> Progressive cultures have worked to free humanity from the brutalities of mere survival and to reduce suffering, improve society, and develop global community. We seek to minimize the inequities of circumstance and ability, and we support a just distribution of nature's resources and the fruits of human effort so that as many as possible can enjoy a good life.

Humanists are concerned for the well-being of all, are committed to diversity, and respect those of differing yet humane views. We work to uphold the equal enjoyment of human rights and civil liberties in an open, secular society and maintain it is a civic duty to participate in the democratic process and a planetary duty to protect nature's integrity, diversity, and beauty in a secure, sustainable manner.

Thus engaged in the flow of life, we aspire to this vision with the informed conviction that humanity has the ability to progress toward its highest ideals. The responsibility for our lives and the kind of world in which we live is ours and ours alone.

Appendix: Humanist and Secular Resources

Join Greg Epstein online at:

www.harvardhumanist.org

www.newsweek.washingtonpost.com

www.thenewhumanism.org

Humanist Celebrations (e.g., Weddings, Funerals, Baby Namings)

THE HUMANIST SOCIETY

www.humanist-society.org

The Humanist Society (HS) prepares Humanist celebrants to lead ceremonial observances across the United States and worldwide. Celebrants provide an alternative to traditional religious weddings, memorial services, and other life events. HS is also starting to work at the grassroots level by developing the Humanist Community Project, providing the tools to develop truly Humanist communities. These communities will empower local Humanist groups with the ability to reach out to a greater population of people by providing educational programs, regular communal celebrations, and meaningful social interaction with fellow Humanists. Visit our Web site to find a Humanist celebrant near you, or to learn how to become a celebrant.

Additional Ceremony Resources

For additional resources regarding Humanist ceremonies, see the Association of Humanistic Rabbis (www.shj.org/AHR.htm). The British Humanist Association also has a wealth of resources (www.humanism.org.uk/ceremonies), including books such as *Funerals Without God: A Practical Guide to Non-Religious Funeral Ceremonies, New Arrivals: A Practical Guide to Non-Religious Baby Namings,* and *Sharing the Future: A Practical Guide to Non-Religious Wedding and Affirmation Ceremonies* (all by Jane Wynne Willson).

Ideological/Community Organizations

American Humanist Association

www.americanhumanist.org
The American Humanist Association (AHA) is busy working through courts, legislatures, media, and local organizations to increase public awareness and acceptance of humanists and like-minded freethinkers. Convinced that we need reason and progressive ideals, not blind faith or backward dogma, at the center of America's thinking, we've been emphasizing growth and outreach like never before. That's why we launched the Kochhar Humanist Education Center last May to support the growing number of adult and youth educational programs rooted in Humanist values. It's why we acquired the Appignani Bioethics Center to raise awareness of bioethical issues from a uniquely Humanist and scientific perspective. It's why we are pursuing a number of legal actions through our expanding Appignani Humanist Legal Center. And it's why we are raising a clear voice for Humanism through high-profile advertising campaigns across the United States.

International Humanist and Ethical Union

www.iheu.org

The International Humanist and Ethical Union is the world union of over a hundred Humanist, rationalist, secular, Ethical Culture, atheist, and free thought organizations in more than forty countries. Its mission is to represent and support the global Humanist movement, with the ultimate goal of building a Humanist world in which human rights are respected and all can live a life of dignity. As a federation of national and regional Humanist groups, the IHEU coordinates activities of its member organizations, stimulates their policies and guides their strategies, fosters the growth of new Humanist groups, and represents the interests of Humanists at the United Nations and other international forums.

The Society for Humanistic Judaism

www.shj.org

The Society for Humanistic Judaism advances a human-centered approach to Jewish identity that combines the celebration of Jewish culture with rational thinking and an adherence to Humanistic values and ideas. Established in 1969 by Rabbi Sherwin Wine, the society reaches out to cultural Jews who seek to blend a celebration of Judaism with a nontheistic philosophy of life. The society creates an inclusive, nurturing environment for families with children, empty nesters, preschoolers, teens, university students, young adults, seniors, single parents, the GLBT community, and intercultural families. By organizing congregations and creating celebrations for Shabbat, Jewish holidays, and life cycles, by publishing resources and educational materials, by offering adult and children's programs in communities and through HuJews, its national youth organization (www.hujews.org), and by supporting the training of rabbis and leaders, the society provides a meaningful alternative for Humanistic Jews.

The society's affiliated communities and congregations offer a Jewish home for many secular Jews who would not otherwise remain part of the Jewish community. Executive director M. Bonnie Cousens and Rabbi Miriam Jerris, Ph.D., the professional staff, are dedicated to building unorthodox congregations for today, while creating continuity for tomorrow.

Center for Inquiry/Council for Secular Humanism

www.centerforinquiry.net
www.secularhumanism.org
The mission of the Center for Inquiry (CFI) is to foster a secular society based on science, reason, freedom of inquiry, and Humanist values. CFI fulfills its mission through a wide variety of programs, including adult education, academic research, legal advocacy, publishing, conferences, campus outreach, and social services, including moral education for children and secular parenting courses. CFI has a network of approximately twenty centers and communities throughout North America, as well as numerous campus groups that sponsor programs and activities and where like-minded individuals can meet and share experiences. CFI also sponsors special research projects, such as The Jesus Project, in which leading scholars are carrying out a multiyear assessment of the evidence for the historical Jesus. CFI also fulfills its mission by providing logistical and technical support for affiliated organizations, including the Council for Secular Humanism. The council publishes *Free Inquiry* magazine and the *Secular Humanist Bulletin* and supports a wide range of activities to meet the needs of people who find meaning and value in life without gods. The council also has about twenty local affiliates, but unlike CFI centers and communities, the local affiliates of the council are autonomous groups.

American Ethical Union

www.aeu.org
The American Ethical Union is a fellowship of people who seek clarification of the values of life and a faith to live by. We cherish freedom of the mind and freedom of conscience. Our affirmation is the worth and dignity and possibilities of every person. Our common ground is concern with the relation of human beings to one another. Ethical Culture is a Humanistic religious and educational movement inspired by the ideal that the supreme aim of human life is working to create a more humane society. There are Ethical Culture Societies in many cities across the United States.

The HUUmanists Association

www.huumanists.org

The HUUmanists Association represents Humanists and Humanism within the Unitarian Universalist Association of Congregations, to whose 1963 General Assembly in Chicago it traces its formal origins. Formerly known as the Fellowship of Religious Humanists, it has evolved beyond its original purpose of promoting a largely academic vision of nontheistic, naturalistic Humanism in a social context closely resembling that of traditional theistic religions. Today the HUUmanists Association, with nearly a thousand members, lay and clerical, is a rapidly growing grassroots network aimed primarily at reaching Humanists and others in UUA congregations everywhere, in a united effort to strengthen Unitarian Universalism by strengthening UU Humanism. Through programming and presentations at the Unitarian Universalist Association General Assembly, regional and local meetings, the distribution of Humanist books, and the publication of a newsletter and a journal, *Religious Humanism*—a forum in print for ongoing conversations and collaboration between Humanist and other religious viewpoints within the UUA—the HUUmanists Association supports the aspirations of individual religious Humanists, and also provides program support to dozens of independent discussion groups, forums, and local Humanist communities with UU connections.

Freedom From Religion Foundation

www.ffrf.org

The Freedom From Religion Foundation has been working since 1978 to promote free thought and to keep church and state separate. The foundation promotes freedom from religion through a weekly national radio show, a newspaper, a free thought billboard campaign, and other educational endeavors, including scholarships for freethinking students. The foundation acts on countless violations of the separation of church and state, and has taken and won many significant complaints and important lawsuits to end church-and-state entanglements and challenge faith-based initiatives.

Atheist Alliance International

www.atheistalliance.org

The Atheist Alliance International is a worldwide organization of democratic atheist associations and individuals, including over sixty groups across five continents. A nonprofit membership organization, AAI supports the establishment of democratic atheistic societies and fosters their success and growth for the purpose of advancing the atheist worldview and protecting atheist rights. In addition to sponsoring annual conventions, AAI publishes the quarterly magazine *Secular Nation, The Freethought Directory*, and the *Journal of Higher Criticism*. AAI represents more than five thousand members internationally and supports its members with information, networking opportunities, and guidance as they face the challenges of living and promoting the atheist worldview.

American Atheists

www.atheists.org

Since 1963, American Atheists has been laboring for the civil liberties of atheists and the total, absolute separation of government and religion. It was born out of a court case begun in 1959 by the Murray family that challenged prayer recitation in the public schools. Now in its fourth decade, American Atheists continues working for the civil rights of atheists, promoting separation of church and state, and providing information about atheism.

The Congress of Secular Jewish Organizations

www.csjo.org

The Congress of Secular Jewish Organizations (CSJO) is composed of independent organizations that promote a secular expression of its members' Jewish heritage, with particular emphasis on the culture and ethics of the Jewish people. For us the continuity and survival of the Jewish people are paramount. The prophetic tradition of social justice and Humanism is the foundation upon which our continuity is built. Our schools and adult and youth groups function outside the framework of organized religion and carry out programs of education directed toward understanding our people's past and enriching our present Jewish lives. These programs include study of our tradition, history, literature, music, art, and languages (Yiddish as a vital instrument of expression of a significant period of our history; Hebrew as it relates to modern Israel; and other Jewish languages created in the Diaspora). Creative approaches to holiday celebrations provide an opportunity to reflect upon our cultural and historic heritage and to relate their significance to present-day life.

Special Interest and Support Organizations

Secular Coalition for America

www.secular.org

The Secular Coalition for America is a political and cultural advocacy organization whose purpose is to amplify the diverse and growing voice of the nontheistic community in the United States. It is located in Washington, D.C., for ready access to government, activist partners, and the media. Its members are established national nonprofits that serve atheists, agnostics, Humanists, freethinkers, and other nontheistic Americans. Their purpose in founding the coalition in 2002 was to formalize a structure for visible, unified activism to improve the civic situation of citizens with a naturalistic worldview. The coalition is the first such cooperative venture in the history of the American free thought movement. A primary focus of the Secular Coalition for America is direct lobbying of Congress and the executive branch with the aims of: 1. presenting secular Americans as a diverse, growing constituency and political force; and 2. promoting public policy that preserves the separation of religion and government and protects the rights and interests of nontheistic Americans. The coalition also facilitates grassroots lobbying through its action alert system, which enables participants to contact their elected officials from the coalition's Web site.

Secular Student Alliance

www.secularstudents.org

The mission of the Secular Student Alliance (SSA) is to organize, unite, educate, and serve students and student communities that promote the ideals of scientific and critical inquiry, democracy, secularism, and human-based ethics. SSA supports a network of over 140 atheist, agnostic, Humanist, and other nontheistic groups on high school and college campuses around the world. It provides these groups with speakers, grants, promotional materials, staff support, and an annual conference.

United Coalition of Reason

www.unitedcor.org

The United Coalition of Reason exists to raise the visibility and sense of unity among local groups in the community of reason and to create a national dialogue on the role and perception of nontheists in American society. It carries out this work primarily by providing funding and expertise to help local groups cooperate toward the goal of raising their public profiles. Specifically, United CoR acts as a catalyst for the forming of informal local coalitions of atheist, free thought, Humanist, nontheistic religious, skeptical, and other similarly minded organizations. It then offers free Web hosting and a design template for the local Coalition of Reason Web site, provides free public relations and media training for local group leaders, and provides funding for a local publicity campaign aimed at bringing the public to the CoR's Web site. Since the central focus of United CoR is local, it doesn't compete with other national organizations, and its local work is geared toward fostering the success of existing groups, not changing their nature or adding new groups or an additional organizational level to the mix. United CoR is also independently funded and therefore doesn't compete for support with national and local entities in the community of reason it serves.

International Institute for Secular Humanistic Judaism (IISHJ)

www.iishj.org

Secular Humanistic Judaism is a human-centered philosophy of life combining rational thinking and Jewish culture with the best ethical insights of the Jewish and human tradition. It is a nontheistic celebration of Jewish identity that affirms the power and responsibility of individuals to shape their own lives. The International Institute for Secular Humanistic Judaism (IISHJ) is an academic and intellectual center for cultural, secular, and Humanistic Jews. The IISHJ provides general adult education through seminars, colloquia, and publications. It also offers professional training for Secular and Humanistic Jewish spokespersons, educators, leaders, and rabbis in North America. The institute includes faculty members of major universities throughout the world who serve as part-time lecturers and instructors. Distinguished writers, intellectuals, and ordained Secular Humanistic rabbis also serve as faculty. The initial IISHJ office is in the Pivnick Center for Humanistic Judaism in Farmington Hills, Michigan; the IISHJ administrative center is currently in suburban Chicago.

Foundation Beyond Belief

www.FoundationBeyondBelief.org

Foundation Beyond Belief serves two purposes: to focus, encourage, and demonstrate Humanist generosity, and to support Humanist parent education programs. Foundation members choose their preferred level of automatic monthly donations to featured charitable causes in nine categories—health, hunger, human rights, peace, child welfare, environment, education, animal welfare, and small charities—and set up an online profile to distribute their contributions among the causes as they wish. Online forums provide an opportunity to discuss the foundation's work and the place of Humanism in philanthropy today. Our featured charities are not limited to explicitly Humanist organizations but to the fulfillment of the central value of Humanism—caring for each other and for this life and this world. They are carefully selected for their impact and efficiency and may be founded on any worldview so long as they do not engage in proselytizing. The foundation also engages in leadership training and programmatic support for nonreligious parenting workshops nationwide under the direction of Dale McGowan, author and editor of *Parenting Beyond Belief* and *Raising Freethinkers*. After a prelaunch phase in late 2009, the foundation's vibrant online presence will fully launch in January 2010.

Skeptics Society

www.skeptic.com

The Skeptics Society is a scientific and educational organization dedicated to the promotion of science and critical thinking. The society engages in scientific investigation and journalistic research to investigate claims made by scientists, historians, and controversial figures on a wide range of subjects. The society also engages in discussions with leading experts in our areas of exploration. It is our hope that our efforts go a long way in promoting critical thinking and lifelong inquisitiveness in all individuals. Some people believe that skepticism is the rejection of new ideas, or worse, they confuse skeptics with cynics and think that skeptics are a bunch of grumpy curmudgeons unwilling to accept any claim that challenges the status quo. This is wrong. Skepticism is a provisional approach to claims. It is the application of reason to any and all ideas—no sacred cows allowed. In other words, skepticism is a method, not a position. Ideally, skeptics do not go into an investigation closed to the possibility that a phenomenon might be real or that a claim might be true. When we say we are skeptical, we mean that we must see compelling evidence before we believe.

Atheist Nexus

www.atheistnexus.com

Atheist Nexus is an online community for atheists, agnostics, Brights, freethinkers, Humanists, and skeptics. Forums and social networking tools allow members to meet other nontheists and discuss a wide range of topics in a supportive environment.

FreeThoughtAction

www.freethoughtaction.org

FreeThoughtAction (FTA) is an independent marketing campaign designed to help the splintered free thought movement speak with one voice to increase the visibility and influence of the large and growing nontheist community. FTA's "Don't believe in God? You are not alone" billboard first appeared in January 2008 on the New Jersey Turnpike and has since been installed in nine additional markets as of this writing, with more on the way. The campaign has helped coalitions of varied free thought groups come together in common purpose in several cities and has been instrumental in the recent launch of the United Coalition of Reason (United CoR), which is taking the campaign to dozens of additional cities across the country. The billboards and related activities have generated tremendous media coverage, including stories in the Associated Press, *New York Times*, Fox News Channel, and *Playboy Magazine* as well as many local news outlets in the markets in which they have appeared.

Camp Quest

www.camp-quest.org
Camp Quest is a network of summer camps for the children of atheists, agnostics, freethinkers, Humanists, Brights, skeptics, or whatever other terms might be applied to those who hold to a naturalistic, nonsupernatural worldview. The purpose of Camp Quest is to provide children of freethinking parents with a residential summer camp dedicated to improving the human condition through rational inquiry, critical and creative thinking, scientific method, self-respect, ethics, competency, democracy, free speech, and the separation of religion and government. Camp Quest fulfills this mission through providing a mix of hands-on educational activities and traditional summer camp fun, including canoeing, arts and crafts, swimming, and games. Camp Quest serves campers ages eight through seventeen, and some locations provide family camps for parents and younger children as well. Although Camp Quest is aimed at children of freethinkers, all children in the appropriate age ranges are welcome to attend. Adults age eighteen and older can volunteer as camp counselors. Currently eight Camp Quest camps hold sessions in Ohio, Tennessee (Smoky Mountains), Minnesota, Ontario (Canada), Michigan, California, Florida, and the United Kingdom.

Military Association of Atheists and Freethinkers

www.maaf.info
The Military Association of Atheists and Freethinkers is a community support network that connects military members from around the world with

each other and with local organizations. In addition to our community services, we take action to educate and train both the military and civilian community about atheism in the military and the issues that face us. Where necessary, MAAF identifies, examines, and responds to insensitive practices that illegally promote religion over nonreligion within the military or unethically discriminate against minority religions or differing beliefs. MAAF supports constitutional separation of church and state and First Amendment rights for all service members. We also coordinate with other national organizations that hold the same values.

Humanist Institute

www.humanistinstitute.org

Are you now or are you interested in being a Humanist: community leader, UU minister, chapter leader, Ethical Culture leader, national nonprofit leader, educator, counselor, or leader in a serving profession? The Humanist Institute exists to equip Humanists to become effective leaders in a variety of organizational settings, including within the Humanist movement itself. The institute is an independent graduate-level program that works in cooperation with existing Humanist organizations. It provides a unique opportunity to bring together a diverse faculty and student body of those who take a nontheistic, naturalistic approach to Humanism, whether interpreted in secular or religious terms. We strive to prepare Humanist leaders to be advocates who are persuasive articulators of the Humanist worldview in the public sphere, innovative thinkers who keep our movement alive with provocative insights and writings, and experts who have the skills to help organizations succeed.

James Randi Educational Foundation

www.randi.org

The James Randi Educational Foundation is a not-for-profit organization founded in 1996. Its aim is to promote critical thinking by reaching out to the public and media with reliable information about paranormal and supernatural ideas so widespread in our society today. The foundation's goals include:

Demonstrating to the public and the media, through educational seminars, the consequences of accepting paranormal and supernatural claims without questioning.

Supporting and conducting research into paranormal claims through well-designed experiments utilizing the scientific method and by publishing the findings in the JREF official newsletter, *Swift*, and other periodicals.

Assisting those who are being attacked as a result of their investigations and criticism of people who make paranormal claims, by maintaining a legal defense fund available to assist these individuals. To raise public awareness of these issues, the foundation offers a prize of $1 million to any person or persons who can demonstrate any psychic, supernatural, or paranormal ability of any kind under mutually agreed-upon scientific conditions. This prize money is held in a special account that cannot be accessed for any purpose other than the awarding of the prize.

Acknowledgments

This book is dedicated to Sherwin T. Wine, without whom it could not have been written. But it could just as easily have been dedicated to my mother, Judy Capel, with whom I began having a fascinating conversation on many of these issues not long after I was born. Her encouragement of and curiosity about my work have been an inspiration.

Tom Ferrick gave me the chance to follow him in this wonderful career, and Dr. Joe Gerstein has made my path infinitely easier and more meaningful. Siwatu Moore has been a true friend, and an enormous moral support.

My agent, Robert Guinsler, has shepherded me through the writing process and been a real mensch, as has my very talented editor David Highfill.

And I have been continually inspired and instructed by my community members and students at Harvard, by many of the star Humanist leaders across the country and beyond, and by many others I've been fortunate to know: student and alumni leaders John Figdor, Dan Robinson, Lewis Ward, Amanda Shapiro, Kelly Bodwin, Andrew Maher, Greta Friar, Dave Rand, Brendan Randall, Peter Blake, Sebastian Velez, Jason Miller, Hann-Shuin

Yew, and many more; board members Mike Felsen, Tom Larkin, Tony Proctor, and Marcia Flynn; advisers Jim Anathan, Woody Kaplan, Steve Pinker, and E. O. Wilson; volunteer leaders Rekha Vemireddy, Bob Mack, John Silva, and Jenni Acosta; research asssistants Alissa Ford, Garga Chatterjee, Sarah Chandonnet, Andrea Runyan, and Jackie Lewandowski; Maureen George; Steve Rade, Fred Edwords, Jende Huang, and the United Coalition of Reason; Roy Speckhardt, Dave Niose, and the American Humanist Association; Bonnie Cousens and Rabbi Miriam Jerris at the Society for Humanistic Judaism; August Brunsman and Hemant Mehta of the Secular Student Alliance; Rabbi Sivan Maas, Professor Yakov Malkin, Felice Posner Malkin, and all my wonderful colleagues at Tmura in Jerusalem; Ambassador John Loeb; Lou Appignani; and Marilyn Rowens.

I'd like to thank the Pivnick and Posen families, who steadfastly supported my education and that of many others on this path; my family—Jon Epstein, Emily Klein, Ruth Seymour, Ruth Behar, the Gordons, the Van Grovers, and Shoshana Levin and family at Kibbutz Ga'ash; Sally Quinn, Jonathan Haidt, Ralph Williams, Rebecca Goldstein, Eboo Patel, R. Lanier Anderson, Michael Ignatieff, Sir Salman Rushdie, Joss Whedon, Greg Graffin, Rep. Pete Stark, Bret Anthony Johnston, Diana Eck, Scott Brewer, Tom Clark, Debra Dawson, my fellow Harvard chaplains, Rev. Professor Peter Gomes, all the SMART Recovery facilitators doing important and unsung work; Duncan Crary, Marc Bernstein at the New York Society for Ethical Culture; my former band mates, Sugar Pill; Rachel Tronstein, Ron Luhur, Chris Farah, all the Boston Frisbee organizers and players; Julie Sheinman and the alums of the Stuyvesant High School Speech and Debate Team; and Machar, the Washington Congregation for Secular Humanistic Judaism—including its rabbi, my dear friend Ben Biber, executive director Roz Seidenstein, and former Sunday school teacher Elizabeth Karcher, for inspiring the "Ten Commandments" section.

Notes

Introduction

1. This statement, admittedly somewhat jarring given that so many other minorities have and continue to face discrimination in politics and beyond, is based on a good deal of data on American voting preferences collected by the Gallup Poll. See for example: http://www.data360.org/report_slides.aspx?Print_Group_Id=99. See also the following study by sociologists from the University of Minnesota, which found that Americans rate atheists below Muslims, recent immigrants, gays and lesbians, and other minority groups in "sharing their vision of American society": Penny Edgell, Joseph Gerteis, and Douglas Hartmann, "Atheists as 'Other': Moral Boundaries and Cultural Membership in American Society," *American Sociological Review* 71 (2006): 211–34.
2. See Barry A. Kosmin and Ariela Keysar, *American Religious Identification Survey 2008, Summary Report* (Hartford, CT: Institute for the Study of Secularism in Society and Culture, Trinity College, 2009).
3. See Phil Zuckerman, *Society Without God: What the Least Religious Nations Can Tell Us About Contentment* (New York: New York University Press, 2008).
4. Joss Whedon, speech for the Humanist Chaplaincy at Harvard, Memorial Church, April 10, 2009.

Chapter 1: Can We Be Good Without God?

1. Rick Warren, *The Purpose Driven Life* (Grand Rapids, MI: Zondervan, 2002), 38.
2. Ibid., 37.
3. C. S. Lewis, *The Abolition of Man* in *The Quotable Lewis*, ed. Jerry Root and Wayne Martindale, pp. 72–80 (Carol Stream, IL: Tyndae House, 1990), 29.
4. Albert Mohler, "Can We Be Good Without God?" http://www.albertmohler.com.
5. Sayyid Qutb, *Milestones*, quoted in Gustav Niebuhr, *Beyond Tolerance: Searching for Interfaith Understanding in America* (New York: Penguin, 2008), 61.
6. Seyyed Hossein Nasr, *The Heart of Islam: Enduring Values for Humanity* (New York: HarperCollins, 2004), 45–46, 220.
7. Adolf Hitler, *Mein Kampf*, trans. R. Manheim (Boston: Houghton Mifflin, 1943), 1:65.
8. Brian Swimme and Thomas Berry, *The Universe Story* (San Franciso: HarperSan-Francisco, 1994), 7.
9. Ibid., 5.
10. Richard Dawkins, *The Blind Watchmaker* (New York: Norton, 1996), 5.
11. Ronald W. Clark, *Einstein: The Life and Times* (New York: Avon, 1984), 502.
12. See Paul Tillich, *The Dynamics of Faith* (New York: HarperCollins Perennial Classics, 2001) or John Shelby Spong, *A New Christianity for a New World* (San Franciso: HarperSanFrancisco, 2001).
13. Tillich, *The Dynamics of Faith*, 4.
14. See "Oprah Winfrey's Commencement Address," Wellesley College, May 30, 1997, http://www.wellesley.edu/PublicAffairs/PAhomepage/winfrey.html.
15. Sherwin Wine, "Reflections," in *A Life of Courage*, ed. Dan Cohn-Sherbok, Harry Cook, and Marilyn Rowens, 284 (Farminton Hills, MI: The International Institute for Secular Humanistic Judaism and Milan Press, 2003).
16. Christopher Hitchens, "The God Hypothesis," in *The Portable Atheist* (New York: Da Capo Press, 2007), 235.
17. See Daniel Dennett, *Breaking the Spell* (New York: Viking, 2006), 199.
18. For the article from which these five rules are taken, see Nowak's "Five Rules for the Evolution of Cooperation," *Science* (December 2006), 1560–63. For a more in-depth look at the evolution of cooperation, see Nowak's *Evolutionary Dynamics* (Cambridge, MA: Harvard University Press, 2006).
19. Nowak, "Five Rules," 1560.
20. Charles Darwin, *The Descent of Man* (New York: Penguin Classics, 2004), 137.
21. Stephen Jay Gould and Richard Lewontin talked about spandrels first in "The Spandrels of San Marco and the Panglossian Paradigm: A Critique of the Adaptationist Programme," *Proceedings of the Royal Society of London, Series B, Biological Sciences* (1979): 205, 581–98.
22. For a more in-depth but still very accessibly written summary of these ideas, see Robin Marantz Henig, "Darwin's God," *New York Times Magazine*, March 4, 2007.
23. Hitler, *Mein Kampf*, ed. Ralph Manheim (New York: Mariner Books, 1999), 152.

24. See David Van Biema, "Mother Teresa's Crisis of Faith," *Time*, August 23, 2007.
25. John F. Haught, *God and the New Atheism: A Critical Response to Dawkins, Harris, and Hitchens* (Louisville, KY: Westminster John Knox Press, 2007). Quote is in an excerpt from the book in an article by Haught, "Amateur Atheists: Why the New Atheism Isn't Serious," *Christian Century*, February 26, 2008.
26. *The Dialogues of Plato*, trans. Benjamin Jowett (New York: Charles Scribner's Sons, 1902), 290.
27. Paul Chamberlain, *Can We Be Good Without God?* (Downer's Grove, IL: Intervarsity Press, 1996), 188.
28. For more on this topic, see the rather creatively titled online forum http://whygodhatesamputees.com.

Chapter 2: A Brief History of Goodness Without God, or a Short Campus Tour of the University of Humanism

1. I'm thankful to my friend and colleague Dr. Adam Chalom, a Humanist rabbi from Deerfield, Illinois, who often teaches the history of Humanism, for pointing out the example from Friedman and coloring it with this Yiddish saying.
2. Wendy Doniger O'Flaherty, *The Rig Veda* (New York: Penguin Books, 1981), 25–26.
3. Madhava Acharya, *Sarva-Darsana-Samgraha*, trans. E. B. Cowell and A. E. Gough (London: Trubner and Co., 1882), 10–11.
4. Ibid., 10.
5. Sarepalli Radhakrishnan and Charles A. Moore, eds., *A Sourcebook in Indian Philosophy* (Princeton, NJ: Princeton University Press, 1973), 233–34.
6. This saying, sometimes known as the Tetrapharmacon, does not appear in the few extant writings of Epicurus, but is an Epicurean formula generally considered to date to Epicurus himself. This translation appears in Gilbert Murray, *Five Stages of Greek Religion* (New York: Doubleday, 1955), 205.
7. Epicurus, "Letter to Menoecus," trans. Robert Drew Hicks, http://classics.mit.edu/Epicurus/menoec.html.
8. Ibid.
9. Ibid.
10. *Mahapurana* 4:16–31, in *Sources of Indian Tradition*, 2nd ed., ed. Ainslie Thomas Embree, Stephen N. Hay, and William Theodore DeBary (New York: Columbia University Press, 1988), 1:80.
11. As my friend Matt Cherry has pointed out in an excellent essay, "Introduction to Humanism: A Primer on the History, Philosophy, and Goals of Humanism," accessible at http://humanisteducation.com/demo.html.
12. Jennifer Michael Hecht, *Doubt: A History* (San Francisco: HarperSanFrancisco, 2004), 216.
13. Ibn Warraq, *Leaving Islam: Apostates Speak Out* (Amherst, NY: Prometheus Books, 2003), 52.
14. F. Gabrielli, "La Zandaqa an 1er Siecle Abbasiole," in *L'Elaboration de l'Islam* (Paris: Presses universitaires de France, 1961).

15. Abu Bakr al-Razi, *The Spiritual Physick*, trans. A. J. Arberry, as cited in Warraq, *Leaving Islam*, 55.
16. Hecht, *Doubt*, 228–29.
17. Thomas Jefferson, letter to William Short. Thomas Jefferson, letter to John Adams, *The Writings of Thomas Jefferson*, collected and edited by Paul Leicester Ford (G.P. Putnam's Sons, 1899), 143.
18. Thomas Jefferson, letter to John Adams, *The Writings of Thomas Jefferson*, collected and edited by Paul Leicester Ford (G.P. Putnam's Sons, 1899), 185.
19. As cited in Brooke Allen, *Moral Minority: Our Skeptical Founding Fathers* (Chicago: Ivan R. Dee, 2006), 96.
20. Charles Darwin, *The Autobiography of Charles Darwin* (New York: D. Appleton and Company, 1893).
21. Frances Wright, *Life, Letters, and Lectures*, as cited in Annie Laurie Gaylor, *Women Without Superstition* (Madison, WI: Freedom From Religion Foundation, 1997), 34.
22. Elizabeth Cady Stanton, as cited in Gaylor, *Women Without Superstition*, 129–30.

Chapter 3: *Why be Good Without a God?* Purpose and *The Plague*

1. Warren, *The Purpose Driven Life*, 27.
2. See Michael Slackman, "Fashion and Faith Meet, on Foreheads of the Pious," *New York Times*, December 18, 2007.
3. Based on Donald A. Crosby, "Nihilism," *Routledge Encyclopedia of Philosophy* (London: Routledge, 1998).
4. Ibid.
5. See the BBC documentary *The Power of Nightmares: The Rise of the Politics of Fear*, written and produced by Adam Curtis (2004).
6. Jonathan Haidt, *The Happiness Hypothesis* (New York: Basic Books, 2006), 219.
7. Daniel Handler, *Adverbs* (New York: HarperCollins, 2006), 19.
8. Stephanie Coontz, *Marriage: A History* (New York: Penguin, 2006), 23.
9. Eva Goldfinger, *Basic Ideas of Secular Humanistic Judaism* (Farmington Hills, MI: International Institute for Secular Humanistic Judaism, 1996), 10, 18.
10. Simone de Beauvoir, *The Ethics of Ambiguity* (New York: Kensington, 1976), 105.
11. Albert Camus, *The Plague* (New York: Vintage, 1991), 306.
12. Sherwin Wine, "Personal Ethics," *Humanistic Judaism* 12, no. 2 (Summer 1984).
13. Erich Fromm, *Man for Himself* (New York: Fawcett Premier, 1965), 249–50.
14. Haidt, *The Happiness Hypothesis*, 238–89.
15. These quotations are from the very good film version with Kenneth Branagh and Robert DeNiro. The lines do not appear in Mary Shelly's original text.
16. Thomas Friedman, "A Poverty of Dignity and a Wealth of Rage," *New York Times*, July 15, 2005.
17. See Jean Paul Sartre, *Existentialism Is a Humanism* (New Haven, CT: Yale University Press, 2007), 353.

18. de Beauvoir, *Ethics of Ambiguity* (New York: Citadel, 2000), 86. All rights reserved. Reprinted by arrangement with Kensington Publishing Corp. www.kensingtonbooks.com.
19. Fromm, *Man for Himself*, 219.
20. Steven Pinker, e-mail message to author, March 7, 2008.

Chapter 4: Good Without God: A How-To Guide to the Ethics of Humanism

1. See Alexander Stille, "Scholars Are Quietly Offering New Theories of the Koran," *New York Times*, March 2, 2002.
2. Lloyd and Mary Morain, *Humanism as the Next Step* (Amherst, NY: Humanist Press, 1998), 1–2.
3. Adapted from an exercise on the commandments taught in the "Jewish Cultural School" Sunday school of Machar, Washington, Congregation for Secular Humanistic Judaism. Find Machar at http://www.machar.org.
4. See Israel Finkelstein and Neil Asher Silberman, *The Bible Unearthed: Archaeology's New Vision of Ancient Israel and the Origin of Its Sacred Texts* (New York: Simon and Schuster, 2002).
5. Yehuda Amichai, "When I Banged My Head On the Door," in *The Selected Poetry of Yehuda Amichai*, rev. ed., ed. and trans. Chana Bloch and Stephen Mitchell (Berkeley: University of California Press, 1996), 118–19.
6. Jonathan Haidt, *The Happiness Hypothesis* (New York: Basic Books, 2006), 167.
7. Anna Dreber, David G. Rand, Drew Fudenberg, and Martin A. Nowak, "Winners Don't Punish," *Nature* 452 (March 20, 2008): 348–51.
8. Amy Sutherland, "What Shamu Taught Me About a Happy Marriage," *New York Times*, June 25, 2006.
9. See "Reluctant Vacationers: Why Americans Work More, Relax Less, Than Europeans," *Knowledge@Wharton*, July 26, 2006.
10. Dr. Joseph Gerstein, e-mail to the author, October 22, 2008.
11. Ibid., October 23, 2008.
12. Tony Judt, "Europe vs. America," *New York Review of Books*, February 10, 2005.
13. Corliss Lamont, *The Philosophy of Humanism* (Washington, DC: Humanist Press, 1997), 225.
14. Erich Fromm, *Man for Himself* (New York: Fawcett Premier, 1965), 27.
15. Steven Pinker, "Ethics: The Moral Instinct," *New York Times Magazine*, January 13, 2008.
16. Alan Dershowitz, *Rights from Wrongs* (New York: Basic Books, 2004), 2.
17. Ibid., 8–9.
18. Richard Gregg, "The Value of Voluntary Simplicity" (Wallington, PA: Pendle Hill, 1936).
19. Duane Elgin, *Voluntary Simplicity: Toward a Way of Life That Is Outwardly Simple, Inwardly Rich*, (New York: Quill, 1993), 30.
20. Gordan Kaufman, *In the Beginning . . . Creativity* (Minneapolis, MN: Fortress Press, 2004), 38.

Chapter 5: Pluralism: Can You Be Good *with* God?

1. Kenneth T. Jackson, "A Colony with a Conscience," *New York Times*, December 27, 2007.
2. Remonstrance of the Inhabitants of the Town of Flushing to Governor Stuyvesant, December 27, 1657. New York Historical Records.
3. Gustav Niebuhr, *Beyond Tolerance* (New York: Viking, 2008), xxxiv.
4. Sarah Vowell, "Radical Love Gets a Holiday," *New York Times*, January 21, 2008.
5. Eboo Patel, "Religious Pluralism in the Public Square," *Debating the Divine*, no. 43 (Washington, DC: Center for American Progress, 2008), 21.
6. Stephen Prothero, *Religious Literacy: What Every American Needs to Know—And Doesn't* (New York: HarperOne, 2008), 14.
7. Ibid., 222.
8. Daniel Dennett, *Breaking the Spell* (New York: Viking, 2006), 327.
9. Niebuhr, *Beyond Tolerance*, xix.
10. These do's and don'ts were originally developed for a workshop I led along with Hemant Mehta, "The Friendly Atheist." Visit his excellent blog, www.friendlyatheist.com.
11. See Jeff German, "Brown Vindicated in Legislative Prayer Battle," *Las Vegas Sun*, June 3, 1997.
12. See the LCCR Web site: http://www.civilrights.org/about/history.html.
13. Boyce Upholt, *Philadelphia City Paper*, August 6, 2008.

Chapter 6: Good Without God in Community: The Heart of Humanism

1. Mason Olds, *American Religious Humanism* (Minneapolis, MN: Fellowship of Religious Humanists, 1996), 185.
2. Corliss Lamont, *The Philosophy of Humanism* (Washington, DC: Humanist Press, 1997), xvi.
3. Ibid., xvii.
4. See *Humanism and Education in East and West: An International Round-Table Discussion Organized by UNESCO* (Paris: UNESCO, 1953).
5. Alister McGrath, *The Twilight of Atheism: The Rise and Fall of Disbelief in the Modern World* (New York: Doubleday, 2004), 192.
6. For examples of such discourse, see Assaf Moghadam, "A Global Resurgence of Religion?" *Weatherhead Center for International Affairs Working Paper No. 03–03* (August 2003) and Rodney Stark, "Secularization, R.I.P. (Rest in Peace)," *Sociology of Religion* (Fall 1999): 249–73.
7. Samuel Huntington, *Who Are We? The Challenges to America's National Identity* (New York: Simon and Schuster, 2004), 337.
8. McGrath, *The Twilight of Atheism*, 264–65.
9. Ibid., 265–66.
10. Cited in Jonathan Haidt, *The Happiness Hypothesis* (New York: Basic Books, 2006), 237.

11. From http://www.smartrecovery.org.
12. Jonathan von Breton, "Introduction to REBT," http://www.smartrecovery.org/resources/library/Articles_and_Essays/Additional_Articles/intro_rebt.htm.
13. Herbert Benson, M.D., with Miriam Z. Klipper, *The Relaxation Response* (New York: William Morrow, 1975), 107.
14. Ibid., xxi.
15. Studying and practicing *The Relaxation Response* together can be an excellent exercise for groups of Humanists and the nonreligious wanting to experience community and connection while promoting unity with something natural yet larger than our individual selves.
16. Robert Hass, *Poet's Choice* (Hopewell, NJ: Ecco Press, 1998), 15.
17. As cited in Haidt, *Happiness Hypothethis*, 193.
18. Ibid.
19. Find one version of this reading, developed over the course of numerous services at Wine's Birmingham Temple, in Sherwin T. Wine, *Celebration: A Ceremonial and Philosophic Guide for Humanists and Humanistic Jews* (Buffalo, NY: Prometheus Press, 1988), 321.
20. David Brooks, "Goodbye George and John," *New York Times*, August 7, 2007.
21. Wine, *Celebration*, 378.
22. This is from a funeral I performed, but it closely follows a suggested template for a Humanist funeral found in *Funerals Without God: A Practical Guide to Non-Religious Funeral Ceremonies* (London: British Humanist Association, 1998), 23.
23. Sir Salman Rushdie, acceptance speech for Lifetime Achievement Award in Cultural Humanism, "New Humanism" Conference, Harvard University, April 21, 2007. http://americanhumanist.org.hnn/archives/index.php?id=295&article=0.
24. Fred Edwords, "Celebrating Our Humanism," keynote address, HumanLight Celebration (Bridgewater, NJ), December 18, 2005. http://humanlight.org/wordpress/perspectives/fred-edwords-2005/.
25. Clifford Geertz, "The Impact of the Concept of Culture on the Concept of Man," *New Views of the Nature of Man*, ed. John R. Platt (Chicago: University of Chicago Press, 1966), 93–118.
26. Russell does acknowledge that in a "geographical sense" he and all his peers remain Christians. For Russell's purposes, at a time when the old powers still dominated the world cultural stage with almost unquestioned authority, this cultural reality could be easily "ignored." But in a postmodern world where civilizations clash, ignoring geographic—or cultural—reality would seem to be about as useful an approach as ignoring the Taliban in pre-9/11 America.
27. Sarah Vowell, "Radical Love Gets a Holiday," *New York Times*, January 21, 2008.
28. See Anthony Pinn, "On Becoming Humanist: A Personal Journey," *Religious Humanism* (Winter/Spring 1998).
29. Ibid.
30. Richard Wright, *Black Boy* (New York: Harper, 1945), 113–15.
31. Pinn, "On Becoming Humanist."
32. Ibid.

33. See Mark Lindley, *The Life and Times of Gora* (Mumbai, India: Popular Prakashan, forthcoming), 27.
34. See Ibn Warraq, *Leaving Islam: Apostates Speak Out* (Amherst, NY: Prometheus Books, 2003). See also Ibn Warraq, *What the Koran Really Says: Language, Text, and Commentary* (Amherst, NY: Prometheus Books, 2002).
35. Rushdie, acceptance speech, "New Humanism" Conference.
36. See Tu Weiming, "The Ecological Turn in New Confucian Humanism: Implications for China and the World" in *Confucian Spirituality*, Volume Two: *World Spirituality*, ed. Tu Weiming and Mary Evelyn Tucker (New York: Crossroad Publishing, 2004).
37. Stephen Batchelor, *Buddhism Without Beliefs: A Contemporary Guide to Awakening* (New York: Riverhead Books, 1997).
38. Sherwin Wine, *A Life of Courage*, ed. Dan Cohn-Sherbok, Harry Cook, and Marilyn Rowens (Farmington Hills: MI: The International Institute for Secular Humanistic Judaism and Milan Press, 2003), 28.
39. Wine, "Reflections," in *A Life of Courage*, 291–92.
40. Howard Radest, *Toward Common Ground* (New York: Frederick Ungar, 1969), 27.
41. Ibid., 28.
42. Ibid., 38.
43. Ibid.
44. Ibid., 86. Not only this fact but the vibrant, thriving British Humanist Association of today ought to give pause to anyone who might suggest that in the developed West, only Americans with their fixation on the church are interested in Humanist community.
45. Ibid., 40.